What Others Are Saying About
Secrecy Order

Opposite worlds and conflicting desires clash in this action-packed, page-turning suspense. If you like David Baldacci and John Grisham, you will like the Marc Wayne series.

Richard Paul Evans
#1 New York Times bestselling author
The Christmas Box
The Walk series
Michael Vey series

P. S. Wells never disappoints. *Secrecy Order* explodes onto the page with thrilling excitement and races through to the end with twists and turns that kept me wondering what would happen next.

DiAnn Mills
Award-winning, bestselling author
Airborne and *Concrete Evidence*

This is an adventure thriller with strong writing by two authors who have what it takes.

James Rubart
Bestselling Author
Rooms and The Five Times I Met Myself

Secrecy Order is top notch. The increase in heart rate from the intrigue and beautifully crafted, movie worthy plot and descriptions are enough to send your pulse rate nuclear.

Deborah McCormick Maxey, PhD
Award-winning Author
The Endling

Secrecy Order takes the reader on an adventure into modern espionage, the struggle for power between nations, and the timeless sway of familial love. I especially recommend it to those fascinated with the ever-changing world of innovation and technology.

Debbie W. Wilson
Give Yourself a Break

PeggySue Wells' writing grabs, squeezes, and pulses my heart.

Christy Lou Lindsay
Author and Audiobook Narrator
Freeing Colt

Secrecy Order

Things Are Not As They Appear

P.S. Wells
Max Garwood

Pegwood Publishing

Title: Secrecy Order
By P.S. Wells and Max Garwood
ISBN: 978-1-7331060-5-4

Subjects: 1. Fiction/Action & Adventure
2. Science/Physics/Magnetism
3. Law/Intellectual Property/Patent

Cover photo: nOahKEaton, 99designs.com

Published by Pegwood Publishing
Roanoke IN 46783

Printed in the United States of America

Dedication
To our readers.

And to fellow authors
James Rubart,
DiAnn Mills,
and
Richard Paul Evans
Who liked and encouraged this story.

P.S. Wells

Prologue

To lure technology thieves who are plundering prime concepts from the United States Patent and Trademark Office, the FBI requested small-town patent attorney and inventor, Marc Wayne submit a patent application with tantalizing military applications. The bait worked too well. Now America's enemies have the plans and the inventor.

Held in an undisclosed laboratory in China, Marc struggles to convert an unproven idea into a weapon of global superiority. If he fails, his sister dies. If he succeeds, he will never see home and family again.

A team of FBI agents, led by young Mallory Wayne, strive to trace the stolen patent only to find its author—her brother—has disappeared. The first-time project leader is cognizant of the danger to her brother's life. So far, the kidnappers remain elusive.

Colonel Jai Yao knows what it is to be hungry. To elevate his nation for the good of the people, he seeks to develop the Meissner Device. When his team cannot make the stolen patent design functional, he forces the inventor—Marc—to complete the machine for him.

Marc teams up with a young physicist, Lei Quong who longs to return to the United States and the man she is engaged to marry. Marc stages an explosion in the lab and the appearance that he and Lei have escaped in the Meissner Device. The two hide themselves inside a shipment of illegal arms due for transport.

Though he has no idea where they will go, they have escaped their captors.

Chapter 1

His box was being pried open. Marc Wayne wasn't certain how long he had been folded inside the small rectangle, but he felt desperate for water. He squirmed, trying to encourage blood flow to his numb feet. Cramped, different parts of his body had taken turns screaming in agony during the transport. Mercifully, a few areas eventually lost feeling due to pinched circulation. Not as courteous, his neck, shoulders, and hips shrieked with pain.

With a final snap, the wooden lid broke open, and Marc squeezed his eyes closed against the sudden rush of blinding sunlight.

"Son of a…" the oath seamlessly streamed into another language that Marc guessed to be Arabic. He recognized *cheekia*, slang for guns. And *dhimmi*, a word meaning Christian set apart for religious tolerance by the Koran.

Someone roughly grabbed his shirtfront and jerked Marc upright. Spasms wracked his legs with searing pain. Dragged out of the box and to his feet, Marc wobbled weak-kneed, needles of circulation returning to his lower legs, and keenly aware of an AK-47 pressed into his ribs.

Behind him, he heard the top pulled off the second box and the lewd cries of men who found lovely young Lei where smuggled weapons should have been. He didn't know the language, but Marc knew the tone in those voices. When he heard Lei cry out in protest, he wheeled and lunged at the swarthy man who had his hands on her breasts.

"Leave her alone." Marc barreled into the handful of men around the standing but weaving Asian girl. Strong arms grabbed Marc and held him. The circle of men laughed as he struggled against them while their comrade eyed the girl. A large, bearded man moved in to stroke the girl's dirty, but smooth cheek.

"Don't touch her." Marc yelled again.

Lei slapped the man's hand. Angered and humiliated in front of the others, the bully quickly grabbed her wrists and spun her around so her back pressed against his chest. Her eyes wide with fear, Lei struggled. He gyrated against her while the others joked with words Marc couldn't translate but still understood.

With his free hand, the bully reached down to the front of his pants.

"No!" Marc screamed against the violation of this college student who had trusted him to get her safely back to the United States and the man she loved. Jerking free from the men who held him, Marc charged Lei's attacker like the Wabash Cannonball. Half the other man's size, Marc's momentum carried him headfirst into the brute's kidney. Off balance, the target toppled and the two sprawled on the unyielding ground where Marc battered the man's temple hoping to even the odds against his superior strength.

A thunderous boot to his ribs sent Marc rolling in the dust. He looked up just in time to see the rifle before the butt crashed against his head.

Chapter 2

In the empty cavernous space, FBI Special Agent Mallory Wayne watched the video again. Her research had located the former United States Patent and Trademark Office (USPTO) employee responsible for the initial leaks of patent information with military significance. This interview confirmed her preliminary theory. Patents vital to national security, like the superior jet engine currently propelling Asian military aircraft, were being compromised inside the Patent and Trademark Office.

Having her conjecture confirmed boosted her tormented self-confidence as a research analyst with the Bureau. The corollary she searched for, yearned for like an addict for the next hit, was concrete information that would lead her to Marc. Today.

As the lead on her first project, Mallory's instructions from her boss were to, "Find the leaks. Shut them down, and while you're at it—quite frankly—get some good PR."

Acid burned in the pit of her stomach. Confident her training and experience were adequate for the challenge, she had been too cavalier anticipating success. Her simple plan had been to submit an application to the USPTO that would be irresistible to a technology thief. Then track the application back to the thief, shut down the leaks, and assure the public the FBI remained hard at work protecting the security of the United States from intruders and bad guys.

She recalled the meeting with FBI coworkers where she laid out her strategy. "We opted to search for an idea with military potential from an inventor completely separate from any government connection. A real person so no one in the information pipeline would get suspicious. Someone with a plausible idea and a legitimate application."

"Does this prince charming come with a glass slipper?" Special Agent in Charge Logan Deverell had stopped clicking his pen incessantly and tucked the pen behind his ear.

"We found a hobbyist who submits patent applications on a fairly regular basis," her associate, Thomas Brenau assured.

Mallory had turned to face Thomas. "It's not a hobby to him."

Thomas put up his hands in surrender. "My apologies, Mallory."

Deverell cleared his throat. "Who does he work for?"

"For himself." Mallory leaned her elbows on the conference table. "And he's a patent attorney, so he would generate his own applications."

"We're not sure if his freelance patent work for other inventors supports his inventions or if his inventions support his office," Thomas noted.

Deverell finished his coffee and, with his practiced technique, tossed the cup at the trashcan. The cup hit the rim and bounced to the floor. "What does this patent attorney and semi-hobbyist inventor have that a foreign entity would be interested in?"

"The invention we selected is for a vehicle propelled by the generation of a Meissner Field." Mallory held her hand, palm side down, and circled it above the tabletop.

"A what? English people," Deverell retrieved the cup and dropped it in the trash, "we speak English here."

From her well-prepared research, Mallory handed a computer printout to her boss. "If this could be developed, the Meissner Device would have substantial commercial and military ramifications."

Deverell had scanned the report. "So, how did you technophobes come up with that?"

"The design was suggested as being the Holy Grail of magnetic inventions by the hobbyist"—Thomas glanced apologetically at Mallory—"inventor."

Deverell scanned the information on the printout. "Knights, a round table, Excalibur, and a love triangle. Go on."

"This particular inventor," Mallory added, "is well known for his inventions centered on electromagnetism."

Deverell looked to Thomas. "Assure him that the FBI simply needs the application to look realistic. The plan is to get his permission to submit and track this patent application. We'll dangle the Holy Grail and see which foreign knight comes courting."

That step, as Mallory expected, had been easier than easy. "We have his permission."

Thomas had swiveled his chair to face her. "You've already told him what we have in mind?"

She nodded.

Thomas directed his attention back to Deverell. "We already have his permission."

"People," Deverell's voice was low, "who is this guy?"

Thomas hooked a thumb in Mallory's direction. "Mallory's brother."

The supervisor's eyebrows shot up. "Brother?"

"Her baby brother."

Deverell looked from Thomas to Mallory. "Your brother is an inventor?"

"And patent attorney," Thomas listed.

"Well," she had hesitated, wondering how personal—how vulnerable—to be. "He's my adopted brother, actually."

"That explains everything." Deverell retrieved the pen from behind his ear and began his annoying clicking.

She forged carefully ahead. "My parents adopted him—"

"I understand that part." He clicked his pen with each word for emphasis. "You're suggesting involving family members. That move proved disastrous in Camelot."

That was the moment, Mallory knew now, when she should have listened to wisdom. Most policies were in place for a reason. Instead, like the amateur she was, she had defended her plan. "I recognize that. But Marc really is an inventor. He is a patent attorney known for submitting his own patent applications. An application filed by Marc will appear natural. Expected."

Ever since Marc had come home as a winsome infant, she had been fiercely protective of her brother. Until the day she compromised. "Rather than raise suspicions," she had continued to sell her strategy, "a submission to the United States Patent and Trademark Office from a respected patent attorney should not scare away our info thief."

Thomas had supported her idea. "He fits the profile."

The boss had sighed and rubbed the back of his neck.

Mallory pressed, "And he's really working on a—"

"Meissner Field Generator." Deverell thumbed through the papers in front of him.

As lead on her first project, she had given her best recommendation—her only recommendation—for moving forward with this case. Waiting for Deverell to approve or reject her suggestion, Mallory forced herself to remain quiet.

Finally, Deverell closed the folder and stood. "All right. Let's run with it." He left the room whistling *Camelot.*

But no one was whistling now. As predicted, Marc's patent application had been stamped with a Secrecy Order. But Mallory had not predicted that the patent office employee, Hatim Saad, who had examined Marc's application for a Messner Field Generator, would be found dead. Next, Marc had disappeared, and all the evidence indicated he had been kidnapped. Whoever had taken her brother didn't play nice, and she was desperate to find him before it was too late.

But what if she was already too late? The guilt she held at bay threatened to envelop her like a tsunami. Mallory pushed that possibility aside. She couldn't dwell there, or all hope would be gone. In a grieving state of mind, she could miss an important clue that could lead her to Marc. Until she had irrefutable proof of the contrary, she chose to believe finding Marc remained a possibility.

In truth, Mallory had never felt so frightened in her life.

Focused and serious, Thomas and Deverell searched for leads. Since the stolen patents appeared to have been compromised at the patent office, they had turned their attention there.

Combing the records of people employed at the USPTO, Mallory noted one examiner who had been employed during the period when applications stamped Top Secret had been leaked to foreign governments.

Retired from USPTO, Arnold Taylor had made major changes in his lifestyle, changes that included a large amount of money in a secondary bank account. Follow the money. The money trail never failed as a source in Mallory's research. Especially with criminals that were not astutely shrewd. Manipulated and used, pompous Taylor deceived himself into imagining himself as some sort of brilliant spy that absconded with the Crown Jewels.

With Hatim Saad dead, Mallory investigated past and present employees that would be likely candidates for Saad's predecessor. Taylor had left a trail for her to follow as clear as footprints in the Indiana snow. As project manager, Deverell fed the lead to field agents who paid a surprise visit to the retiree.

Once she had received and reviewed the video of their interrogation, Mallory sent a text to Deverell asking to meet. He replied he was out of the office on fieldwork, and she would have to rendezvous someplace convenient to his assignment.

Considering what was available near his location, she found a place and sent a message. *Meet me. July 30.*

Now, in an empty coliseum, Mallory viewed the video once more on her phone, concentrating on each word. There had to be additional information she could exploit in her desperate search for her brother. Realizing she had the episode memorized, she got up from the plastic folding seat and found a vending machine. Depositing quarters, she chose a root beer for herself and a diet coke for her boss.

Were her choices a mistake? If Deverell took much longer to arrive, the coke would be warm. She groaned. Terrified she would make another bad choice or send the investigation in the wrong direction and further endanger Marc, she second-guessed her every decision. Hyper-cautious, she could mess this up even more than she already had. She could barely decide what soda to buy.

Walking, she circled the large arena several times. She knew if her thoughts kept ricocheting like a bullet in a submarine, she would be no good to anyone. Not to herself. Not to her team. Certainly not to Marc.

In the upper stands, a door opened. Light and Deverell came into the cement corridor. The previous July, their department had busted jewel thieves at the National Barrel Racing Youth World Competition. Under the guise of cowboy bling, stolen diamonds had been transported in plain sight at the coliseum through sales vendors that lined the upper floor. From the date she texted, he knew where to find her.

Chomping a wad of gum, her boss took the cold soda she offered. "You found something."

Mallory motioned Deverell to a row of seats. She replayed the video on her phone. The spec ops guys had tugged a hood over Taylor's gray head and 'coptered him to the landing pad on a business building and into the hands of agents who pushed him, whining and tripping over his feet, down an elevator and into a small room. Shoved into a metal chair, he quaked when they removed the hood.

Blowing cigarette smoke, a guy whose face looked like he made side money in a boxing ring, leaned menacingly close to the slouched Taylor. The questioning began, and when Taylor spat in his captor's face, the move became his last defiant act. As a red welt rose across Taylor's cheekbone from being backhanded, a second guy laid out rows of photos and copies of bank and credit card statements. Proof that Taylor had been on the take and indulged himself with the payoff of his betrayal. Having Taylor's full attention, the man with the paperwork waved a badge under the former patent office worker's panicked gaze.

Mallory tapped the screen. "This is the traitor that stole and sold the patent for the GE jet engine."

"To the Chinese?" Deverell popped the top and gulped the drink.

"Not directly."

"A middleman?"

She nodded.

"What else did we learn?"

"Shocked at all we knew about him, he began to comprehend the depth of the trouble he was in. The word espionage loosened his tongue. The news that his replacement had been killed, and that he could be next, motivated Taylor. He became all too eager to share what he knew in exchange for protection."

"The emperor's new clothes are a sham, and our boys informed the emperor that he is naked." Deverell tipped back his head and finished the last of his diet coke. "What do you have so far?"

"We know Marc was taken by ambulance from his office in Dixon, Indiana, to Smith Field in Fort Wayne and airlifted to the west coast in a medical flight. We've tracked the pilot, and the accompanying doctor—if you can call him that. He's wearing an eye patch and under investigation with the medical licensing board." She pointed to the man on the video. "Now, we've taken into custody the previous USPTO employee leak."

"What did you get from him?"

Mallory pocketed her cell phone. "We got a description of the guy who develops the contacts in the Patent Office."

Deverell crushed the empty aluminum can. "Guys like Hatim?"

She nodded. "Guys like Taylor and Hatim Saad."

Chapter 3

"Yala, yala," demanded a man's voice from somewhere nearby. "What have you found?"

Marc groaned and rubbed his head. He winced when his fingers found a large goose egg that had risen along the side of his face.

"These two crates do not carry the weapons." A heavily accented voice gave the report. "Instead, we found these two stowaways."

For the second time, muscled arms grabbed Marc and jerked him to his feet. His head pounded, and his vision swam. Squinting, Marc searched for Lei. The man with the lustful eyes that Marc had charged held her arms, but she appeared unmolested. Marc sagged with relief and turned his attention on the newcomer who stood in front of him.

Younger than Marc, this olive-skinned man wore clothing like the Bedouins around him, only cleaner. Intelligent eyes regarded Marc though his words addressed the others. "So, you decided to rape the girl and then see who these people are and what they can tell you about your missing weapons?"

The others shifted uncomfortably.

"Bachir," the man ordered. "Release her."

"But …"

The newcomer spit empty sunflower shells on the bigger man's shoes. With a look, he silenced the opposition. "Where will she go, my friend? Can she outrun your jeeps in this desert? Even your surefooted donkey?"

Released, Lei adjusted her clothes and came to Marc's side.

"Forgive us," the man said, "for not extending traditional desert hospitality to our guests." His words generated hurried movement among several men. "I am Adi."

Marc shook off the unwanted grip of his captors. "Marc. And this is Lei."

A breeze blew from behind them, and Adi wrinkled his nose. "You have been in these shipments for a while?"

"We are sorely in need of washing." Marc brushed at his shirt.

"Come." Adi turned and led the way.

Following their long-legged host, Marc eyed their surroundings. They were on a hillside overshadowed by higher hills fringing a canyon. Below, the rocky terrain flowed into a wide, dry wadi bed. Two donkeys, a camel, a rust-fringed Mitsubishi truck, and a dusty jeep were parked in the shadow of the hill they were making their way down. The threesome followed a shallow horizontal groove carved into the thirsty hillside. Gradually angling down, the man-made depression disappeared under the base of a rock-lined well.

Perched on the edge of the well, Adi tossed a bucket into the water below. Hand over hand, he pulled the rope that brought the filled container back to the top. Balancing the bucket on the rim, he removed the fabric tied about his head. He strained the water through his scarf into the trough of hewn rock that sat beside the well. Three times he repeated this process. With a wave of his hand, he indicated they should help themselves.

Marc put a hand on Lei's shoulder. "Go ahead."

Her cheeks still aflame, she bent to splash water on her face, and he saw she trembled. Adi noticed too and moved several paces away to allow them some privacy.

"I'm sorry." Marc spoke quietly. "Are you all right?"

She threw another handful of water on her face to camouflage the tears that streamed down her cheeks. "Apes."

"Ugly apes."

Lei splashed water across her arms, rubbed her hands, and sat heavily on the well where she breathed deeply.

Marc washed his own face, neck, and arms, grimacing when the water stung the side of his face where the goon had struck him with the butt of the rifle. Noting the hole in his shirt earned during their escape, he cleaned the dried blood that had caked around his shoulder wound. Sudden images of their flight from captivity exploded in his memory. Firing the M72 LAW to blow a hole in the Chinese development lab, the sickening sound of crushing bone when he swung the missile launcher like a Babe Ruth baseball bat and connected with the guard whose bullet felt like fire when it grazed Marc's shoulder. Eluding those who imprisoned them by securing Lei and himself inside crates of weapons slated to be shipped to who knew where.

He glanced at the severe terrain. They had been shipped here, to this desert with crude men who spoke their own languages. Marc eyed Lei as she studied their surroundings and wondered if he had brought them out of confinement and into a worst danger. Could he get them home to the United States? Could he find and free his sister from the captors who held her?

Adi approached, and they followed him to a colorful carpet someone had spread nearby. He motioned for them to sit. Their host sat cross-legged facing them. Dishes of hummus, flat bread, cheese, olives, and oranges were set between the three, along with bottled water and hot Turkish coffee. Eagerly, Marc and Lei dug into the food.

"Who are you, and why are you here?" Adi waited with his question until they had finished eating.

Marc took a long drink of water. "I'd like to ask you the same thing."

Adi regarded them, his expression kind compared to the first men they had encountered. "What is that between you and me? Since you are enjoying my hospitality, I invite you to answer first."

Chapter 4

Mallory entered the digital forensics department. The men and women who worked in this division were a breed apart. Actually, they often reminded Mallory of her equally unique brother, Marc. Academically brilliant, this group wasn't prone to invest much time or salary shopping the latest clothing fashions, but they easily wore classic styles of inimitable Birkenstocks and khaki pants. Ball caps displayed the department's spirited rivalry over sports teams.

At Jimmy's workstation, the hyper computer genius spastically moved back and forth between two computers, typing on each keyboard with a swiftness that sounded like a machine gun with a silencer. At her arrival, Jimmy put up his index finger, indicating he needed another couple minutes to complete his project.

Mallory held a finger to her lips as Deverell burst into the room. Oblivious, her supervisor planted himself in front of the concentrating Jimmy. "What do you have for us, Jimmy?"

The computer forensics expert ignored the interruption, giving a final tap on the keyboard with the flourish of a composer bringing an orchestra concert to a theatrical end. He swiveled his chair and acknowledged the two spectators.

"Look at this." Jimmy pulled an earbud from one ear. "Steganography algorithms by definition have a vast variety of applications. So far, I've identified over 600."

Deverell whistled through his teeth.

Classic Rolling Stones played from the earbud before he turned off the muted tunes. "But they all work on basically the same principle."

Mallory leaned closer. "Which is?"

"To embed a message in an image." Jimmy pushed his John Lennon glasses further up his nose.

Mallory felt the flutter of hope. Perhaps Jimmy had found something that would lead her to Marc. Following Hatim Saad's death, the FBI studied everything that had crossed the USPTO examiner's desk. Combing through the mail and his phone records, the most interesting discovery showed up in the computer's history. A Hallmark card of a kitten stood out as completely out of place among Saad's records. Magnified, the image displayed background noise.

"We've identified the extra strokes as extra pixels." Jimmy explained. "Steganography was used by Greek generals who tattooed sensitive information onto the shaved heads of messengers. Once their hair grew back, the messenger traveled without suspicion to the intended recipient who decrypted the message by once again shaving the messenger's head."

"What do bald heads and tattoos have to do with Puss in Boots?" Deverell folded his arms across his chest.

"Using innocuous documents, usually an image file like this cat," Jimmy enlarged the image on the screen, "steganography encodes the message while at the same time concealing the fact that a message is being sent at all."

"Today," Mallory put in, "steganography makes use of email. For any corporate spy or disgruntled employee, this method is an ideal carrier."

Jimmy nodded. "A commonly used steganography algorithm called LSB takes advantage of the way computers digitally encode color. The algorithm hides the fugitive file inside the noncritical bits of color pixels."

"Noncritical? I thought they were all necessary." Deverell loaded cinnamon flavored gum into his mouth.

Jimmy shrugged. "Like croutons on a salad, these are the least important information in a pixel."

"I like croutons," Deverell said.

"Fattening." Jimmy circled a finger in front of the picture. "See the gray in the cat's fur?"

Deverell studied the screen. "Ninety percent of the fur is gray."

"Exactly," Jimmy said. "Every pixel is a number. The gray is coded as a number much like 00 10 01 00. But by changing the least significant bits—in this case the last two—the programmer produces a one-millionth of a color change."

Mallory leaned closer. "That's so absurdly subtle that your eye cannot detect it."

"Go on." Deverell tapped his foot.

"The steganography application weaves the secret message into the least significant bits of the image." The digital forensics expert looked at his audience with expectation. "Understand?"

"Doesn't the alteration show up in the process of sending it?" Mallory listed, "What about compression? Or different formats?"

Jimmy shook his head. "The image file is unaltered in variables such as size, JPEG, in lossy compression or in lossless compression."

"Okay, that's how Hatim delivered the information. How does the receiver unlock the message," Mallory wanted to know.

"The receiver of the message has the original image, and he uses an unlocking algorithm to locate the stowaway bits in the cat image," Jimmy indicated the picture on the second computer's screen, "and uses them to reconstruct the secret message."

Mallory caught herself twirling her hair and covered the nervous tick with a quick scratch on her ear. Jimmy had proven her theory. Patent information vital to national security had been leaked to enemies of the United States. Thanks to the work of this department, now she knew the vehicle used to transfer the information in patents before they could be developed on

American soil. She felt reassured that her analysis of the situation had been right on target.

"Nice work, Jimmy." If only she hadn't used Marc to gain the proof she needed.

Deverell clapped the computer forensics guy on the shoulder. "I need you to find a way to halt this method of transporting our secrets into the hands of enemy governments."

Excitement of the challenge gleamed in Jimmy's eyes. "I'll get right on it."

Deverell folded his arms. "Where has the pussycat been, and what did she do there?"

Chapter 5

While the others went about their business at the desert site, Adi drove Marc and Lei out of the wilderness. The man Adi called Reuven sat in the passenger seat, and Marc saw that he carried a handgun in his waistband.

Though their driver remained calm, Marc found the breakneck speed unnerving as they traveled along cliff edges, plunged into deep dry wadi beds, and climbed steep passes in the dusty vehicle. The few times Adi used the brakes, they piercingly squealed their protest at being coaxed to work.

At last, an oasis appeared in the desert, and the vehicle sped toward this place. As they drove, the desert gave way to acres of carefully maintained orchards, gardens tented under white plastic, and pasture animals grazing among sparse crops of wild grasses. Adi parked the jeep beside a cluster of well-kept buildings.

"Where are we?" Lei's legs were wobbly after the wild ride. Before Marc could offer, Adi took her hand to steady the girl as she climbed from the jeep.

Their host gave a grand sweep with his free hand. "This is my home, my kibbutz."

"Kibbutz?" Lei frowned. "I don't understand."

Adi looked to Marc who also shook his head.

The rusty truck parked behind them, and three armed men followed Adi at a distance.

Trailed by the brooding Reuven, Adi led the way. "In my country—"

"I assume we are in Israel," Marc interrupted.

The younger man nodded. "In Israel, some 117,300 people live in kibbutzim."

"How many kibbutz?"

Lei's question allowed Marc to scope the surroundings and take in whatever information he could gather at their urgent pace. He assembled two lists in his head, one of possible materials and the other of people who may or may not be helpful in his quest to get to Mallory.

"From the Golan Heights in the north," Adi indicated a direction, "to the Red Sea in the south, there are 268 such communities. Some have less than 100 people, most have several hundred, and in a few cases, there are over 1000 residents in the kibbutz."

Marc surveyed what appeared to be a modern hotel and stylish restaurant. Shops and guest cottages circled an adjacent building, and from the people coming and going in swimsuits, he gathered there must be an indoor pool.

Being in a pocket of civilization meant phones. The ability to make calls. The thought jolted Marc. Who should he contact? Who would be able to help him find Mallory? Could he reach the people she worked with at the FBI? Were they already searching for his sister? Their co-worker?

Marc tuned back into the conversation as Adi described, "Most of the kibbutzim, 80 percent, were founded before the establishment of the State of Israel in 1948."

"Who were the founders?"

Adi walked beside Lei as he explained. "A century ago, a group of young Jewish immigrants came here from Eastern Europe. Inspired by a mixture of Zionist and socialist ideals, they established the first kyutza on the shores of the Sea of Galilee."

Marc half-listened. He needed to know all he could about his new predicament. He also had to help his sister.

"Kyutza?" Lei tried the new word.

"Hebrew for group," their host continued. "As membership grew, the name was changed to kibbutz."

"Which means?"

"Community."

Longing to be back in Indiana, Marc considered how differently the Midwest farms appeared compared to the Middle Eastern crops at this place. "Do you support yourselves through farming?" Was Mallory still in Indiana?

"Initially, yes." Their guide slowed his long strides to accommodate Lei who still moved stiffly after her time in the crate. "Young Jewish pioneers acquired land by the Jewish National Fund. They reclaimed the soil of their ancient homeland and forged a new way of life. Inexperienced with physical labor and lacking agricultural knowledge, they faced a desolate land neglected for centuries. Water was scarce. Funds were scarcer. Against these challenges, they worked hard and developed thriving communities that played a dominant role in the establishment and growth of the fledgling state."

They arrived at a building. The entryway smelled of water and humidity. Behind glass walls on the left was a workout room populated with modern exercise equipment. A physical therapist coached a teen through strengthening motions. The girl's bald head told Marc she was recovering from cancer treatment.

On the right, three shallow pools of differing sizes lay serene under thin clouds of steam that rose from their surfaces. In the nearest pool, an instructor led a class in cardiovascular training. The chests of several participants bore the tell-tale scar of heart surgery.

"Later generations convinced our parents that farming was not enough in modern society. Each kibbutz established a business as well. One makes shoes that are sold worldwide, another developed a drip irrigation system. One provides seeds and specialized green houses, another telecommunication, another software, and yet another medical equipment."

"Yours is tourism?" Marc noted that of the men that shadowed Adi, only the scowling Reuven followed them inside.

"Very lucrative when our enemies are not raining missiles on us from the sky." He indicated the pools. "Guests from many countries come to rejuvenate in the minerals and waters."

Marc glanced sidelong at Adi's companion. "It didn't look like you were booking tourists in the desert when you found us."

As he expected, his comment generated a reaction. Reuven's hand shot out and gripped Marc around the neck. "You know nothing about what we do." His voice was low, and Marc read the threat in the brooding man's eyes.

"Relax, cousin," Adi addressed.

"He asks too many questions." Reuven locked eyes with Marc but spoke to Adi.

"As I would expect you to do in similar situations."

The man sneered a silent warning at Marc before releasing his hold.

Marc rubbed his neck. "At the end of the day, do you throw him raw meat?"

Adi laughed and slung an arm around his cousin's shoulders. "He's a good man, this one."

His cousin easily shrugged off Adi's arm and glowered at Marc. "They threw me to the wolves when I was born, and I returned leading the pack."

Chapter 6

Pow. Pow. Despite the earplugs, Mallory could hear the muffled pop of the Glock as she squeezed off each round. Keeping her focus on the target in the firing range, she ignored the group of tourists behind the sound-proof glass. Agent hopefuls and FBI wannabes led the tours that included the museum in the lobby.

Prior to 9/11, congressmen and senators provided arrangements for citizens in their constituency to visit the FBI building. Now, those rare opportunities were available only to VIPs. Visitors wanted to do three things at the Hoover Building; ask what the FBI did, view items once belonging to high-profile criminals brought to justice, and watch someone fire rounds at the indoor shooting range.

She could imagine the answer to the first question. "Spies. Terrorists. Hackers. Pedophiles. Mobsters. Gang leaders and serial killers. We investigate them all, and many more besides." In the display room, the group could see John Dillinger's bullet proof vest he was not wearing the lethal night agents waylaid him as he exited a Chicago theater, as well as his death mask molded from his face as he lay in the morgue. Much to her chagrin, she was fulfilling the third expectation.

But Mallory wasn't here for anyone else. Not even to practice her skills. Frustrated, Mallory just wanted to blow something to Mars. She could shoot something or cry. And she wasn't going to cry. Certainly not where anyone would see. With explosive frustration as her companion, she decimated a paper silhouette.

Parking the Glock on the shelf, she picked up a .45, a standard for military spec ops. Resettling the plastic earmuffs that served as hearing protection, she gripped the gun with both hands. Focused on the front sight, she squeezed off two shots in succession—a double tap. She fired until the slide locked back on the pistol. She removed the empty magazine and flipped the wall switch. A wire trolley sped the paper target to her. Slightly low and left of center, seven holes peppered the silhouette.

Reloading, Mallory purposely avoided thinking through the list of leads the agency had gathered in their search for her brother. In their driveway basketball court back home, while playing one on one with her, Marc used to suddenly stop.

Annoyed that her brother wasn't engaged in the game, she took advantage of the opportunity to bounce the ball off his head. "Okay, genius. What bolt of inspiration just hit you?"

He would look at her, that light of discovery in his eyes. Marc had learned that when he reached an impasse in the problem-solving process, the best solution involved doing something completely unrelated. Coincidently, while he rested and refreshed, the answer often appeared. Unbidden. Unforced, and of its own accord, like a child peering from a hiding place.

She switched from shooting at a stationary target to the faster pace of aiming at a moving silhouette. Her adrenalin kicked in as the target rapidly came to her like an assailant. Her reaction had to be jaguar fast. Especially with an audience of tourists watching. She didn't want to miss while being observed by friends and network connections of the D.C. elite.

Network connections. Network. Connections. Like being hit on the head by a basketball, the solution appeared. Now she knew where to look for the next clue as she hunted the path that she hoped would lead her to Marc.

Adjusting her ear protection, she aimed at the next target and mentally went over the sequence of events.

During the short window of time immediately following Marc's disappearance, she and Thomas had discovered a sketchy trail. As usual, Marc had ridden his bike early in the morning to his patent attorney office in the small rural town of Dixon in Indiana.

His secretary, Violet Seiwert, had arrived at her normal hour to find the front door unlocked, and Marc's bike parked inside like every other day when he biked downtown. Though the veterinarian who occupied the adjacent office and doubled as the building's landlord said he had spoken with Marc early that morning, the patent attorney was nowhere to be found when Mallory stopped in just after Violet's arrival.

Mallory's purposeful tour of the town including a return home produced no brother. Back at Marc's office, Mallory had pretended to be on her cell phone as she waved apologetically at a worried Violet, pointed to the phone, and went straight back to Marc's office.

His laptop was not in the usual place on his desk. Not a good sign. While many people brought home their computer to work after hours, Marc never did. He knew when to say enough, lock the office door behind him, and turn his attention elsewhere. He told her that his best ideas for inventions came during periods when he gave his brain time off. His backpack carried a notepad, and myriad pieces and parts he often toyed with at home, but not his computer. Customarily, the laptop remained at the patent attorney office for correspondence and patent applications.

Stuffing down an urge to panic, she had hurried to his workroom. Looking carefully, she noticed an empty spot on his cluttered workbench where he stacked his papers scribbled with notes, formulas, and math equations. A thin coat of dust covered the tools and materials scattered to the edges of the workbench. The most used area of the tabletop near Marc's chair had been cleared.

Finding the back door unlocked, she went outside. Two faint lines were visible in the manicured patch of grass leading to the gravel driveway. From her training, she recognized the marks made by heels when someone was dragged. Where the short trail ended, she could see where accelerating dual rear tires had piled dirt.

Her brother's notes were gone. His laptop was not where it should be. And there was evidence that someone was dragged from his office and transported away. A mechanic reported the ambulance scheduled for routine maintenance had disappeared from the parking lot, but when police stopped by, the vehicle had magically returned. Not only did the timing fit for someone to have taken Marc from his office in the ambulance, but no one would have questioned the presence of the emergency vehicle. Considering the amount of time the ambulance had been missing gave Mallory and Thomas a radius that included the nearby airport.

At Smith Field, Mallory had spoken with a pilot who flew medical transports. The morning of Marc's disappearance, an ambulance had delivered a doctor, an Asian male nurse, and patient for a flight to Southern California. Mallory knew they were talking about her brother when the co-pilot mentioned the patient had a ponytail.

Deverell had notified the authorities to be on the lookout for Marc.

Back in her office, she went over the facts with Deverell and Thomas.

"Based on the results of previous leaks from the Patent Office," Deverell had considered, "the patent and our inventor are probably in Asia."

"Great." Mallory had thrown up her hands. "That narrows it down to a mere quarter of the globe."

Thomas crossed to the map on the wall. "Since the cutting-edge jet engine was developed in China, why not begin there?"

At last, a place to begin. "I'll get right on it."

"Whoa." Deverell put a restraining hand on Mallory's arm. "Before you invade the People's Republic, while China developed the engine, based on our information, I don't credit them with the ability to directly acquire top secret material."

Mallory folded her arms. "Then where do we begin?"

"If China is the recipient of the information, they probably received the designs from another entity. Someone that pilfers marketable secrets."

"A governmental entity or private entrepreneur?"

"That's the question. Who has Cinderella's slipper? Has Marc been delivered to the highest bidder, or is he making the Meissner Device operational for the marketers?"

Later, Mallory read through the reports sent in from the Coast Guard, airports, and border patrol, seeking any clue to Marc's captors and his whereabouts. DHU Aerocomposites was an Asian company with suspicious dealings. The same morning Marc had disappeared, their cargo ship, Shanghai Sumi had sailed from a section of the Southern California harbor owned by Chinese shipping conglomerates.

The hulking cargo ship had been searched by the Coast Guard. Though the report said nothing was found, the boarding team may have caught something on their helmet camera that was not recognized. Mallory would review the video feed from the Coast Guard boarding team. Perhaps she would notice something that had been overlooked.

Chapter 7

The next morning, Lei rose early. She had enjoyed a solid sleep-in fresh clothing, stretched out on crisp sheets until early this morning when nightmares about the lecherous Bachir's hands on her body jarred her awake.

Marc had been in a heated argument with Adi last night. While Marc insisted, he had to immediately make contact with the United States, Adi calmly repeated that his guest must wait. All would happen in due time. Lei felt anxious that they had traded one controlled situation for a new one, and Marc accused Adi of exactly that, when a woman named Esther took Lei's arm and led her away from the men who were deciding her fate.

Esther brought Lei to this cottage consisting of a large bath, and an inviting bedroom and sitting area. Lei noted that, while not being obvious, one of Adi's men stayed always nearby.

An extended soak in a fragrant tub imbued with Salty Sea minerals had been followed by a generous application of luxurious soap and soothing lotions. Clean and sleepy, Lei had come out of the bathroom to find Marc laying on the couch, his back to her. She hoped he finally drifted into some much overdue rest. Seeing he was showered, in clean clothes, and his shoulder wound had been properly bandaged, she covered him with a blanket and climbed into bed.

Now, in the moments just after dawn, she stepped outside quietly, so as to not disturb Marc's sleep. Curious, she wanted to see more of this new place and put yesterday's fright behind. There was something exotic and ancient in this land. Like an old soul. Despite the dangers they had faced the day before, she felt drawn to know more of this strange country. Close to her own age, Adi captured her imagination. He embodied an authority even those older desert ruffians respected.

Several others were already moving about the kibbutz. Cheerfully calling greetings, men and women made their way to the gardens and greenhouses. Young women with energetic children talked and laughed as they went into the dining hall, the building where she and Marc had eaten their fill of fresh vegetables, succulent beef, and sweet challah bread last night.

In a sitting area among the paths, not far from her room, Lei spotted Adi. Like Adi's gloomy shadow, his cousin was there, as well as the stately Esther. Seeing her now from this distance, she shared Adi's handsome features. Or he had hers since she was older. These three were in deep conversation with several older men. Lei turned her steps in the opposite direction.

Swooping among the bushes and branches, birds sang birdsong greetings to the rising sun. Lei heard someone whistle an echo to the song of the birds. She bent to smell a flower. Returning to her path, she was startled to find a man standing in front of her.

"Good morning." Adi plucked a bloom. "If I may?" He gently tucked the flower behind her ear, considered his handiwork, and smiled. "You do the flower a great service."

Lei bobbed her head. "Thank you."

"Would you walk with me?"

Feeling shy, she nodded and fell into step beside the graceful man, observing the activity of the kibbutz as the gem in the desert rose with the sun.

"How are you this day?"

"Much improved over yesterday." She felt her face flame as the memory of that horrible man's hands on her body flooded back. Quickly, she pushed away the accompanying fear by changing the topic. "Though, I have many questions."

"Perhaps I have some answers."

"The people here appear healthy and happy."

"Our community is based on common ownership of the means of production, consumption, and education." Adi spoke the words like a catechism. "Conferring together, we make decisions by majority vote, and each bears responsibility for all."

"Is this communism?"

Adi shook his head. "A kibbutz is a voluntary collective built around self-labor, equality, and cooperation."

"Voluntary?" This didn't make sense when she and Marc were kept here against their desire to immediately return to the United States. Admittedly, this felt different from their treatment under the People's Republic. And she knew Israel to be a free country. But why was liberty withheld from her and from Marc?

"Completely. People often come to a kibbutz to rest, and never leave. Children born here are educated at university—"

"Like you?"

He returned the greeting of a pretty young woman who blushed and hurried on her way. "I attended university sponsored by the kibbutz, but I was not born here. My mother was. She left for a while. We returned when I was a boy. After graduating university, members may bring their education back for the good of the community, or they may go into the world. Like my mother, many return."

Lei considered the simple structures around her. "There is a peace here. A quiet joy."

Adi smiled. "That is what attracts others to kibbutzim. Instead of private wealth, each member is responsible for the needs of the members and their families. All possessions are generally owned."

A jeep drove past and disappeared into the surrounding wilderness. Lei looked longingly after the disappearing vehicle. "All possessions are generally owned," she echoed and faced Adi. "Then we can use a vehicle to reach a city and …"

Adi put up a hand. "As I explained to Marc, that may not be the safest action for you."

Lei's shoulders slumped. "But you said …" she struggled to make clear her need.

"I said each member is responsible for the needs of the members and their families. You are my guest, and I will do the same for you."

"But what does that mean for me? For Marc?" Eager to get to the United States, to a familiar and safe place, she didn't bother to conceal her frustration. Was the young American man who had proposed waiting in the Midwest to marry her? And while he had not confided her story, she knew Marc was desperate to return to his sister. "Are we free to go? Is being here voluntary, or are Marc and I captives?"

He pointed to a stone bench and reluctantly she sat. Adi straddled the bench, so he faced her. "It means I will see to your safety and your well-being. It means you may have to be patient."

She gazed off into the distance, calculating the chances of either stealing a vehicle or walking out of this place. Adi must have read her thoughts.

"Lei," his voice was like velvet over steel. "We are a long way from a city. This kibbutz is the deepest in the Negev. The desert is the large heart of my country. Our closest neighbor is the Salty Sea."

When she frowned, he clarified. "Outsiders call it the Dead Sea."

She looked toward the rugged cliffs beyond the gently tended fields. He followed her gaze and pointed to a high overhang, distant yet close enough that she could see two wild goats frisking their way to the top. Balanced on tiny hooves, the smallish animals gained the peak by way of the gradual rise on the backside of the half mountain. Now perched on the edge and silhouetted in the morning light, the goats overlooked the vast canyon far below and rimmed on the opposite side by abrupt crags.

Adi turned her attention to a place behind the goats. At first, Lei could only see rocks. Then a stealthy movement caught her eye. A mountain lion also watched the goats. Slinking swiftly behind cover of boulders, the predator narrowed the distance to its prey.

"They're trapped." Lei shuddered. Perhaps one would escape while the lion took down its companion. "The lion has cut off their route to escape back the way they came."

"So, it appears."

Closer, the lion crouched, preparing to spring. Then the two stepped over the edge. But rather than tumbling head over hooves into the abyss, one followed the other down an invisible path. Clinging to the side of the seemingly impassable precipice, the surefooted goats casually wove a zigzag pattern that took them to safety. Cheated from her breakfast, the lion swung a claw over the edge but could only watch the animals disappear from view. With a cry that echoed off the crags, the big cat returned to the hunt.

On their way here, the jeep had bounced over rocky paths invisible to Lei. Like rainwater following the unmarked trail from a mountaintop until reaching the cistern in the low country, Adi had steered the vehicle through dry wadis, around steep turns, and past small hillside patches of grass where wild deer-like creatures grazed. She realized that Adi knew the desert like those cunning goats.

"I would counsel you to have a knowledgeable guide if you left here." He touched her chin and turned her face to his. "The desert is a dangerous and unforgiving place. But worse than that, I believe there are dangerous men looking for you. I am more concerned about them than our wilderness."

Chapter 8

Adi led Lei to the dining hall. "Most kibbutzim are similarly laid out." Along the way, he described the buildings they passed. "Communal facilities such as the dining hall, auditorium, offices, and library are at the center, ringed by members' homes and gardens. Sports and educational facilities are beyond these, and industrial buildings and agricultural land make up the perimeter."

Inside the hall, delicious aromas and warmth thawed the early morning chill from Lei's body. Their plates piled with colorful variety, people sat in groups and chatted, some in English, many in Hebrew. A few sat alone with a book as their breakfast partner.

Buffet style fish, cheese, hummus, fruit, and pickled eggs were abundantly heaped on trays. Hot and cold cereal were available next to juice and coffee. Adi handed a warm dish to Lei and stepped aside to allow her to go ahead of him. As she made her selections, she noted curious glances from those in the hall. Though she had dark hair like theirs, she looked nothing like these Jewish people. At least at college in the States, there were several other Asian students attending. Even a professor or two.

Her plate filled with persimmons, dates, eggs, slices of firm white cheese, and small smoked fish, she allowed Adi to choose their table near a window.

Suddenly, a commotion at the building's entrance interrupted the serene morning. Everyone's attention focused on the unusual scene, and Lei's fork froze halfway to her mouth. In a jumble of arms and legs, Marc was pushed into the room.

Chapter 9

Colonel Jai Yao's superiors were not pleased. They invested time, manpower, and finances in this project. Following the successful acquisition of the new jet engine, he had been tasked to develop the Meissner Effect Device. Despite securing the talents of his top engineers, including the lovely Lei Quong, his team couldn't turn the patent into reality. In much the same way as he gained possession of the patent for the invention, Yao brought the inventor to his development lab.

His superiors gave him the moon and expected him to produce the sun in return. But there were no results. No military superiority. The entire episode had turned out disastrously, and they blamed him.

The blame was not the issue. That was characteristic. Every debacle required a scapegoat. Assign fault to another lower on the food chain, mete out humiliating and destructive punishment, and a vacuum opened for the next ambitious comrade to do the impossible—to sell their soul to please the unappeasable.

Yao lit a thin Cuban cigar and poured a second shot of tequila. Night closed in, and he didn't bother to chase away the darkness by turning on a light. Tipping back his head he swallowed the fiery liquid. He slammed down the shot glass which landed on the tray that held his mail, rolled off the table, and clanged to the floor.

The mail. Neglected for days. Yao fingered the stack of envelopes and recognized the familiar thickness of his father's usual communication. Suddenly, he yearned for the comfort and innocence of the small village where he had grown up. Shut off from the world, the microcosm of culture felt contained and controlled. Separated from the global playing field, the problems of that community centered around the base concern of staying alive. Their needs were for food, clothing, and shelter. Simple. Easy.

Sighing, he turned on the lamp. In the circular glow of the electric bulb, he slid an ivory opener along the envelope's edge.

The shaky penmanship told of seasonal events in the farming village, news about a new baby recently born, and updates on the health of Yao's mother and father. "The medicine you kindly provide has not arrived. Perhaps you can see to this? Not for me, of course, but your mother is in agonizing pain."

Yao swore. Rheumatoid arthritis had anchored in his mother's body after her brutal beating. Yao recalled his years growing up as a peasant farmer's son in a poor farming village situated in China's countryside. Each year, a traveling preacher arrived. Thin and gentle, the preacher was nameless except that the villagers called him Brother.

"Brother is here," the villagers had whispered as they passed each other in the fields or carried water.

After dark, many neighbors quietly crowded into his parents' home. Yao fell asleep in his mother's lap as Brother spoke late into the night. In the morning, Yao would wake on his own pallet where his father had carried him sometime in the middle of the night. He rose and accompanied his father to the fields, feeling sleepy and always hungry. There was rarely enough food to quiet the growing boy's empty stomach.

As he grew older, Yao sensed an unspoken understanding between the common farmers. The villagers were careful to keep the Brother's brief presence secret from the stern village cadre.

Thinking of his father's crippled hands and his mother's arthritic body, Yao pictured the day he and his father were bent over, working side by side in the green rice fields. Smelling dank and promising, the earth had to be coaxed, even seduced to produce enough crops to see them through until the next year's harvest. Suddenly, they heard screaming. Racing from the fields, Yao saw strangers throwing the few household belongings from Yao's simple home. One man held a worn portion of a Bible over his head with one hand, and in his other hand, he cruelly clutched Yao's mother by her hair, holding her head at an unnatural angle.

"This is what happens to criminals." Superiority in his voice, he yelled to the neighbors who cringed a fair distance away.

Bolting forward, Yao quickly outdistanced his father. Head down, he barreled into the man who held his mother.

"No, Jai," cried his mother as she was thrown to the ground. The man and the teen rolled and sprawled apart. Yao sprang to his feet, ready to fight. From behind, the cadre and the other stranger grabbed him and pinned his arms. A hand gripped Yao's hair in a tight fist, forcing the boy to look as Yao's mother and father were brutally beaten.

"Stop!" Again and again, he screamed the same word. Struggling fiercely against the iron grasp that held him, his body became drenched with sweat and his own tears. He wailed until he had no more voice. At last, the attackers exhausted their rage.

Curled in a fetal position on the bloody ground, Yao's mother did not move. The men who held his father released their hold on the battered and bruised man who crumbled to the ground like a pile of broken sticks. As the men yelled something to the horrified villagers gathered around this nightmare spectacle, Yao watched his father crawl to his wife, place a hand against her face, and sag into unconsciousness.

Those who held Yao threw him aside. A booted foot crushed his fingers, and he felt the bones snap like dry twigs.

The boy's childhood home was set ablaze along with the family's belongings, heaped into a pile like so much rubbish. And then the strangers were gone.

Surprisingly, his parents had recovered, but they bore the constant physical reminders of the brutality of the strangers. His father's hands were stiff and difficult to move. A nasty and debilitating condition wracked his mother's body and caused her to be unable to tolerate soft human touch or even the feel of her bed and blankets against her skin. Unless Yao could get the medicine to her that soothed the disease like coaxing a vicious dog to slumber.

Punishing him for the American's escape was predictable. Now, Yao understood the depth of his superior's displeasure and how cruel and far reaching their castigation. His parents would be denied vital medicine as punishment for Yao's failure.

Chapter 10

Two men brought Marc to Adi. Lei recognized them as part of the cluster Adi spoke with earlier this morning. Close behind, an older man followed. Lei watched Adi for signs of anger. But there were none as he stirred milk into his coffee.

The group reached their table. Adi waved Marc to a seat. "Please, join us for breakfast."

Marc stood, defiant.

"Of course. First you must get your meal." He glanced at the older man. "Uncle, help him get his breakfast."

The older man grunted and steered the silent Marc to the buffet.

The fear that rose in Lei's throat made swallowing impossible. What would happen to them now? Marc had done something that displeased the men who surrounded Adi. How would this impact their request for a speedy return to the United States?

At the buffet, Lei noted Marc neither took nor filled a plate. Uncle did the task, and the other two men once more guided Marc to their table. Uncle's strong hand to Marc's shoulder dropped Marc into the seat next to Lei, and he set the plate onto the table in front of the glowering inventor.

Adi spoke to the older man. "Thank you, Uncle."

Uncle nodded and disappeared out the door with the other two close behind.

Pushing his empty dish aside, Adi folded his hands on the table. "I assume you were borrowing one of our vehicles."

Marc scowled. "Until your goon stopped me."

"Actually, Uncle is one of your countrymen."

"American?" Marc's eyebrows shot up.

"Indeed." Adi sipped his coffee. "He fought in your Vietnam war. After the horrors he experienced there, he came here to rest and heal his soul."

"But he is still here," Lei said.

"For special reasons." Adi met Lei's eyes. "He fell in love with a beautiful Jewish girl. When he returned home to tell his parents he planned to convert to Judaism to marry, his mother wept uncontrollably."

"She didn't want to lose her son." Lei thought of the American boy she loved. Her father had been willing to let her follow her heart to the other side of the globe. Knowing he might see his only child rarely, that his grandchildren would be far away, her noble father had defied the control of his government and given her freedom to make her own choices. When she fell in love with a fellow college student at the Midwest University where she majored in electro-physics, her father traveled to meet the young man, see Lei in her Western world, and gave his blessing for her marriage.

Adi continued his story. "Like you, Uncle thought his mother preferred he remain near his family, until his mother brought out an aged photo of a young woman holding an infant." He paused, and Lei found herself leaning forward. "'Your mother,' she said, 'asked me to take good care of you when she put you into my arms in Germany. You're not converting' she explained. 'You're Jewish.'"

"He belongs here." Lei glanced around at the families gathered for breakfast. She didn't have a place where she belonged. When she received news her father was ill, she immediately returned to China. But he had died before she could say good-bye. China cut off her communication with the United States and put her to work in the lab where her education could be used to serve her country. One day, Marc had appeared in the expansive lab that had once been a temple. She had been surprised to see the inventor who had been a guest lecturer at her college, and when she approached, he had been rude. Later, she looked up to find him at her workstation.

"What university did you attend?"

"Indiana," she stated. "Why?"

"I was trying to figure out where our paths had crossed before." He put out his hand. "I'm Marc Wayne."

"I know." She ignored his hand. "You were the guest lecturer at my university."

"Please accept my apology. I was rude."

She glanced at his outstretched hand and met his gaze. "Yes, you were." She studied him before putting her small palm into his.

He gave her hand a gentle shake, and she was suddenly aware of the comfort of a friendly touch. "And your name, physics major?"

"Lei. Lei Quong. Electro-Physics."

He turned to go and stopped. "What was the topic? Of my lecture?"

"You are being rude again." As she turned away, Marc quickly stepped toward her. "Humor me. It's an honest question."

Her tone had been sarcastic as she waved her hand to encompass the large lab which had become almost a prison, keeping her from returning to life and love in the United States. "The Meissner Effect; Laboratory Oddity—"

Looking stricken, Marc finished for her. "Or Revolutionary Tool for Mankind."

Now Lei turned her attention back to Adi's story. "Uncle belongs here."

Adi nodded. "So, Uncle has come home, and we are grateful for his talents in our community."

"Some talents." Marc rolled his shoulder. "Guerrilla warfare tactics."

"I'm not surprised that you were seeking a way to escape."

"You were expecting as much," Marc groused.

Adi indicated Marc's plate. "You will feel better after you eat."

"I'll feel better when I'm back home."

"Like the harvest, all in due time."

With his fork, Marc poked scrambled eggs. "Where's the ham and bacon?"

"According to Scripture, we don't mix the meat of an animal with its milk. Therefore, we serve milk and cheeses for breakfast. In the evening, we have meats without dairy products." He leaned back in his chair, at peace in contrast to Marc's agitation.

Lei sympathized with Marc's drive to be back in control of his life and future. She wondered if Adi ever wrestled with the same frustration.

Something beyond his desire to get home troubled Marc. His sleep last night had been restless. He had called out to an invisible enemy.

Marc flopped a small fish onto Adi's plate. "This is meat."

"Fish and eggs are neutral and served with meat or dairy. But miniature cows are a completely different matter."

"Miniature cows are called veal back home." Marc emphasized the last word.

"A scripturally clean animal, cows are legal in Israel," Adi said.

"Pigs are quite legal in the U.S."

"Our national law forbids a pig's foot to touch Israeli soil."

"So, no ham or bacon like we have freely in the States," Marc said.

"A sister kibbutz does a large tourist business. In response to the dietary requests of their customers, the kibbutz built an enclosed barn with a wooden floor several feet off the ground. Pigs are raised in that barn without ever touching Israeli soil."

"But they're still pigs," Marc said.

"They're referred to as miniature cows."

"A rose is a rose," Marc quoted, pushing around a persimmon with his fork.

Adi smiled. "And a pig is a pig."

"Does the government crack down on them? Or can they do what they want with their own property?" Stressing the last four words, Marc's eyes challenged the man across the table.

"Some things do not appear expedient at first." Adi met Lei's gaze. "The land in Israel belongs to the government. We rent but cannot own the ground because of what is happening in the United States."

Marc stared at his plate. "Much of the United States' land is owned by foreigners, even foreign governments."

Their host finished his coffee. "We are a new nation and a small one. We cannot afford for any piece to belong to foreigners, especially those hostile to Israel. But we are also a people long experienced in negotiation and compromise. If we have miniature cows whose feet never touch Israeli dirt, what is that between us and our government?"

"The letter of the law rather than the spirit of the law," Marc said around a mouthful of cheerios.

Uncle returned and waited near the entrance. Lei considered the gray-haired man who had come to the kibbutz seeking rest for his soul and had found his history, family, love, and belonging. She hoped her journey would have the same satisfying ending.

"I have work." Adi stood. "Uncle will show you the grounds. Once you have seen the layout, you will understand that it would be foolish, probably deadly for you to attempt again to leave here to find your way on your own through our wilderness." He locked eyes with Marc. "Please understand, the people who seek you are as deadly as our desert. And unlike our desert, they are evil."

Chapter 11

Keeping her expectant anticipation at bay, Mallory viewed the footage of the Coast Guard boarding of Shanghai Sumi. Perhaps Captain Gennett's report proved accurate, and there had been nothing suspicious on the boat.

She hoped she could use their eyes to see something they had missed. Well-respected in the service, Gennett caught the attention of his superiors and traveled a fast-track to a stellar military career. Currently ranked first in the Coast Guard fleet, Gennett's boat achieved top scores in ship inspections and personnel consistently placed high in skills. When other boats proved to be poor performers, Gennett's crew often got bumped to take the assignment. That reputation weighed against Mallory finding anything, but she was determined to explore any possibility.

Equipped with video cameras on their helmets, the boarding team divided and began a well-rehearsed and systematic inspection of the ship. Mallory watched each video recording captured by these trained servicemen, running the video in slow motion to enable her to observe the ship and the crew within range of the camera lens. The process was tedious. Her neck ached, and her back hurt. She rolled her shoulders in an attempt to release the pent-up emotions that had settled as a chronic case of tense muscles.

Time slipped by, and she kept at the task. She needed a breadcrumb in the search for her brother to mark the trail. He had disappeared in September when the leaves fell from the trees and pooled against the sidewalk curbs along her hometown streets in Dixon, Indiana. Now, the calendar announced the arrival of October. A week was too short for a beach vacation but tortuously long when her brother's life was at risk. Mallory knew that with every day that passed, the likelihood of locating him reduced exponentially.

With the coast guardsman, she traveled into the cabin section of the ship. One after another, the cabins were sparse. Until this one.

She sat forward and concentrated on the segment of a cabin with a casket. Marc could be inside, but that wasn't possible, or the Coast Guard search team would have found him. As she watched, the lieutenant was unable to open the sealed casket. She debated this fact. Possibly the seal protected the contents during the extensive transportation. Maybe hindered curious crewmembers.

Maybe this scenario hid contraband—like Marc.

Chapter 12

Uncle's tour included the residential area of simple homes and colorful gardens, children's classrooms and playgrounds, and communal facilities including an auditorium, library, swimming pool, tennis court, medical clinic, laundry, and grocery.

Adjacent to the living quarters were sheds for dairy cattle and modern chicken coops. A short tractor ride away stretched agricultural fields, orchards, and fishponds.

"The desert blooms under your care," Lei said.

Marc remained quiet, and she suspected he worked on another half-baked plan to get them away from the kibbutz and closer to their destination. Since their arrival here, he appeared sullen and driven. Having heard him call his sister's name in his fitful sleep last night, she suspected his thoughts were occupied with getting them to America sooner rather than later.

When they first spoke at the lab, Marc said he missed Mallory. Naturally. But after her own experience, Lei was no longer naïve about the manipulations Yao and his people were capable of to accomplish their ambitions. Using and threatening beloved family relationships proved to be powerful motivations.

Uncle was speaking, pulling her attention away from her questions about Marc and back to the old man. "For the founders, tilling the soil of their ancient homeland and transforming city dwellers into farmers was an ideology, not just a livelihood. Kibbutz farmers coaxed barren lands to produce crops, orchards, poultry, dairy, and fish farming. Organic agriculture is a mainstay of our economy. A combination of hard work and advanced farming methods account for a large percentage of Israel's agricultural output to this day."

To get from place to place within the kibbutz, people walked or rode bicycles. The elderly and disabled traveled via electric carts.

"From each according to his ability," Uncle explained. "And to each according to his needs. That is the philosophy we live by."

Beyond the upscale hotel, they entered the large spa. Separate from the rehabilitation complex they had visited upon their arrival, the steamy facility hosted tourists and guests. "When the kibbutzim could no longer support ourselves strictly through agricultural pursuits, each kibbutz developed a supplement," their guide described. "This is ours."

Visitors experienced mud baths and a series of pools with varying degrees of heat, minerals, and salt content from the Salty Sea. Lei's attention was captured by an elderly couple that painstakingly helped one another down shallow steps into the nearest pool, undulating with pulsing currents. Settling into the therapeutic water, they rested in submerged seating, and massage jets rhythmically soothed stiff joints.

Lei had embarked on this questionable venture, so she could share the seasons of her life and grow old with the young man she had fallen in love with while attending college in America. As they made wedding plans, Lei was summoned back to China because her father was ill. Her heartbreak over the loss of her beloved parent turned to devastation when the Peoples Republic denied her return to America and assigned her to work on a developmental project that could make her nation powerful. When Marc set in motion a daring plan to leave the confinement of the laboratory in Shanghai, Lei quickly packed herself into a shipment of illegal arms.

Passing massage tables, they observed the skilled hands of a trained masseuse as she poured oil onto the scars of a man's back and ministered to the insulted skin. Throughout their walk, Lei noted that Marc was more interested in the physical layout than the philosophy Uncle expounded on. Certain that considering his options occupied Marc's attention, a glance at Uncle confirmed that the old man was completely aware of Marc's thoughts. As were the two men who shadowed them. Yet Uncle had shown them all about the place. While Adi anticipated Marc's determination to get back home with or without help from the kibbutz, he treated Marc and Lei like honored guests. His only request was that Marc be patient.

But Lei knew Marc was beyond being patient anymore.

Chapter 13

The girl haunted his dreams. Yao woke in the night, his bed clothes wet with perspiration. Lustful desire? Or did he really care what had become of her? Where had Lei gone? Had she left with that American in that cursed device?

His thoughts went back to the last time he saw her. Young and lovely, she wore a dress the color of emeralds when he picked her up. He saw the approving glances of his contemporaries when they arrived for National Day's formal festivities. Perhaps a beautiful wife would be beneficial.

Partway through the evening, Lei had become ill. Unable to accompany her home due to his duty, he summoned a taxi. Their brief good-bye became the last time he saw the enticing Lei.

Yao had been enjoying his second gin and the attention of a pretty woman at a formal celebration event for those of his political station when an aide discreetly whispered that something unexplained appeared to be taking place at the lab. The lab where under his direction the Meissner Device was being built. The People's Republic had invested in Yao with the uncompromising expectation that Yao deliver the Meissner Device that could give his nation global dominance.

He had hurried back to the temple, and like countless times before, entered the main door. But unlike other visits, the ancient building's tang of age and incense had been replaced with an odor of battlefield explosives. Outside the lab, medical personnel hoisted an unconscious guard onto a stretcher. Inside, the Colonel stepped over the still body of another guard and impatiently pushed through the stunned workers blocking his path. His gaze went first to the large hole in the wall, then to the empty area at the room's center.

The bodies of two guards lay in unnatural positions, and someone with a medical bag bent over the nearest.

Touching the blackened floor, Yao had rubbed the ash between thumb and middle fingers. He circled the perimeter of the large, burned mark. Last, he examined the hole through the exterior wall. A hole large enough to fly the device through. The others stood silent, waiting.

When his top scientists were unable to bring the stolen patent to reality, Yao had undertaken an audacious plan to bring the inventor from the United States to construct his own patent. Had Marc blasted through the wall to provide an exit for the machine? Where would he go? How far could the Meissner Device travel? Or was there another explanation in the destruction? While working under Yao's observation, Marc Wayne had procured a means of escape. And he took the device with him.

Furiously, he barked orders. Personnel jumped and scurried from the room. Clenching his fists at his sides, Yao marched back to the door where he turned to slowly survey the large laboratory where he had invested his time and influence in recent years. He swore loudly, the angry words echoing in the hollows of the previously bustling center of new technology.

In the hall, men assembled for his instructions. "Divide into teams. Search the building." He glanced at the river through the opening in the far wall. "Search outside. Search everywhere."

He gave orders to notify other government and military personnel to be on the lookout on land and above the ground.

Then he summoned the scientists. They arrived from various points in the city where families were celebrating National Day. Stunned by what they saw, each entered the lab and slowly tiptoed through the debris, trying to make sense of what had happened. Looking like ashamed puppies, they gathered before Yao.

"Study his notes." Disdain dripped from his words, he wanted them to comprehend his disappointment in their lack of achievement. "Find out how he made this work."

Now in the predawn hours when sleep eluded him, Yao recalled every detail he could remember. He left his bed and went outside to the garden. Walking in the moonlight, he thought back to the events that led him to this moment. In the place where he was raised, villagers feared the city folks who descended upon their quiet lives, searching, threatening, bullying. Though his parents' letters never mentioned him, Yao was certain the Brother or another much like him, occasionally visited his childhood village.

From the deep pocket of his robe, he pulled a thin cigar and lighter. He filled his lungs with the biting smoke and blew a cloud toward the stars. The smell reminded him of the times he and his family and neighbors crowded around the hearth fire while Brother spoke to them through the night.

Brother's message had always been the same. To love others as God had proven his love by sending his son as remuneration for the evils of mankind. Grace. Forgiveness. Redemption. To live as the son had shown when he allowed himself to be crucified in the place of sinful men.

As a teen, Yao had ventured into the unknown to provide for his parents. He used his education to improve village living conditions. The Colonel considered himself a good son. His parents' letters said so.

Yet, Brother's message was offensive to the People's Republic. For the remainder of their handicapped lives, his parents lived with the punishment for their faith. And authorities hunted the Brother. If captured, he would be imprisoned. The horrible conditions were designed to penalize the prisoner for thinking or believing outside the prescribed boundaries

The irony was not lost on Jai Yao. The God his parents and the Brother knew expressed gentleness and invited relationship. Their God freely offered gifts of joy, peace, and kindness. The People's Republic suffocated and controlled her own, punishing even thought and belief, suppressing creativity and individuality.

Marc Wayne had ideas he wanted to pursue. Yao had taken the man against his will and forced—threatened—the American to complete his design for the Colonel's ambitions. The Colonel's motivation was to bring improvement to the village of his family. Yet, he knew his parents would be disappointed in his methods. Yao sat on the garden bench and put his head in his hands.

He pictured Lei. Young, smart, and idealistic. There was something sad about her. Of course, Yao knew she had been prevented from returning to the United States. The government commonly manipulated the people for the benefit of the People's Republic. And he had made advances to her. Hinted that a liaison between the two would be a good choice. The arrangement would suit him and give his parents what they longed for. A grandson. They didn't ask for much, and he always found a way to provide for them.

Good. That word kept coming up as he turned over and examined his thoughts and feelings in the predawn hours. Back inside the house, he downed a shot of tequila and poured himself a second. Was good even good enough? The People's Republic did whatever they lusted for in the name of the good of the people. Even though the people had no say in what was good for them. That was decided for the population by an elite few whose lives in no fashion resembled the circumstances of the citizens they made decisions for.

Brutal men had destroyed his home and damaged his family, claiming their actions were for the good of the country. Yao flexed his fingers. The increasing ache in his joints was a keen reminder of the unnecessary damage that had been perpetrated on his hands as a teen.

He recalled the fear of the villagers as they whispered Brother's name. Their desperation as they helplessly stood by while his parents were beaten nearly to death.

The same fear had been in Marc's eyes. Terror Yao had imposed, wielding the power like an invincible weapon to accomplish his own ambitious goals. This time for Yao's own good. He recalled Lei's passive compliance as she did what he required of her. Her work lacked passion. She held no admiration for him and his position, education, or Brass Rat ring. Perceptive, she knew his attention was for his own gratification. Her eyes had the same vacancy as the faces of the people he had known in the repatriation center. If their positions were reversed and the opportunity presented itself, he would seize the moment and run.

Shuddering, he poured tequila and carried the tumbler to his garden bench. Emptying the alcohol into his body, he gave in to the inebriation. In that state, he dared to admit the truth. He had become just like those men who had stolen the lives of his parents. He had become what he most hated—what Marc had called him—a monster.

Chapter 14

Uncle returned Lei and Marc to Adi at his office within a hub of rooms that made up the administration of the kibbutz. Marc reached for the phone that sat on the desk. Instead of the expected dial tone, the sound told him a central switchboard controlled the line.

At that moment, Reuven entered and strode purposely to Adi's side. He spoke low. "It didn't go well."

"What happened?"

With a suspicious glance toward Marc and Lei, Reuven lapsed into Hebrew.

Adi put up his hand to halt the words. "This concerns them."

"But Adi—"

"Cousin, for these two, it is their life."

Reuven frowned. "You are too trusting."

"The key," Adi confessed, "is knowing who and when to trust. Now, tell me in English."

"The buyer expected two more boxes. When he put a knife to the family jewels, Bachir told them there were other goods." Reuven glanced pointedly at Marc and Lei. "Human goods."

"What is that to him?" Taking a cellophane package from his shirt pocket, Adi put sunflower seeds into his mouth.

"The buyer believes there is value. What was in the shipment he claims belongs to him." He tipped his head toward Marc and Lei.

Adi spit sunflower shells. "What is that to us?"

The cousin leaned closer. "Bachir told him to look here, at the Kibbutz."

"The fat pig. He compromised our families." Adi's eyes narrowed. "Where is Bachir now?"

Reuven lowered his voice. "Dead. After he gave them their information, they killed him."

Adi drew himself up straight. "Rather harsh penalty. Killing a supplier is not good for business."

"They said he could not be trusted with this information. Therefore, he could not be trusted to keep his mouth shut about them."

Their host received the information thoughtfully.

"What will you do?"

Adi put several more seeds into his mouth and systematically split them apart with his front teeth.

"Adi?"

"They have value to us as long as they are ours." Adi spit the shells.

Marc opened his mouth to protest this wanton decision of their fate. As he had done with Reuven, Adi put up a hand to halt his words. But Marc was not going to be put off again. He moved closer, and Reuven quickly stepped between Marc and Adi. Wordless, the cousin faced Marc, his fierce expression daring Marc to give him the opportunity to use the fists he clenched at his sides. Marc no longer cared. He had been punched plenty already and had delivered some of his own. He had reached his limit of bullies stalking and controlling him. Marc was going home, and he was going home now. With or without anyone's help.

"They are trouble." Reuven spoke to Adi though his eyes looked directly into Marc's.

Adi put a restraining hand on Reuven's shoulder. "We must hide them until they serve a purpose for us."

Lei reached for Marc, but he shook her off. "We are not here to serve a purpose for you or anyone else." Marc held a stance to deliver his most powerful swing. "We are leaving now, and if anyone tries to stop us, I will kill them."

An unexpected flicker of respect in Reuven's eyes took Marc by surprise. Reuven kept his protective position between this volatile Marc and Adi who appeared to be calculating, weighing, considering. Marc turned and, taking Lei by the hand, started toward the lot where the jeeps were parked.

"I understand," Adi spoke to Marc's retreating back. "And you need to understand. I will get you home."

Marc whirled at the last word. "When? When my hair is as gray as Uncle's? I don't have that kind of time. I have to get back immediately before ..." He stopped. The demon that drove him, the awful fear for his sister—it was too fantastic. How could he explain the unexplainable, and who would believe him?

"If you leave here now," Adi paused, waiting until Marc listened. "The two of you will never make it to the United States."

Angrily, Marc marched back. Reuven moved to place himself between the two as he had moments before, but Adi halted him with a word.

"Are you threatening me?" Marc pointed to Reuven. "Gonna send your goons to hold us here like some sort of desert prisoners?"

Adi shook his head. "I am protecting you."

Marc snorted and turned to go. Taking Lei's arm, he could feel her pulse hammering as he steered them forward. Any moment, Marc expected Reuven's hand on his shoulder. Fueled by his brewing fury, he would relish coming to blows.

Instead of Reuven slugging at him, Adi fell into step beside Marc. "Ruthless men, far worse than the ones who uncrated you, are looking for you. Leave here, and you leave my protection."

Marc kept walking. Lei struggled to keep pace with his purposeful strides.

"These men are at home in the desert. They will quickly surmise your value to your previous captors." He dropped his voice. "Can you protect Lei?"

The final question caused Marc to stop. Flashing through his memory, he saw the raw fear in her face as the dirty men attempted to rape her. Lei would have been horribly hurt if Adi had not shown up when he did. The horror would have been Marc's fault.

By his flex and go efforts, Marc had whisked them away from Colonel Yao's dominion, but he had delivered her directly in the hands of a baser evil. Marc had been helpless, unable to protect her. Just as he had been powerless to shield Mallory. In the harsh light of reality, Marc stood as a hopeless failure. Adi was the real hero.

"We made it this far." Lei took his hands in hers. "We can make it the rest of the way."

"No." Sorrowful, Marc shook his head. "I can't risk you again."

Reuven broke in. "Adi, they are coming here. To take what they believe is theirs. We don't have much time."

Defeated, Marc addressed Adi. "What now?"

"We hide you."

Reuven was ready to get on with things. "Hide them where?"

Adi spit shells at their feet. "In plain sight."

Chapter 15

Mallory raised the volume as high as the sound could go on the Coast Guard video that recorded the boarding and search of Shanghai Sumi. Did Marc lay inside that casket?

In the video, Mallory heard a voice that came from behind the coast guardsman. "Perhaps I can help." The Coast Guard lieutenant turned, and through the camera secured to his helmet, Mallory saw a Chinese man, slightly gray at the temples.

"I want to see inside this casket." The voices were woolen, the quality of homemade family videos.

"Certainly." The Chinese man made a slight bow in the direction of the casket. "The dead is a relative of one of the owners of our company. He traveled to the United States in the vain hope of finding a cure for his disease."

"Disease?"

The older man lowered his voice. "A humiliating condition that caused the flesh to rot from his bones."

"Who are you?"

Mallory repeated the lieutenant's question. Was this the same Asian man the pilot had mentioned? The male nurse aboard the medical transport flight out of Fort Wayne?

The man wore civilian clothes. "The casket, as you experienced, was sealed to prevent not only dishonor to the dead but to prevent the contagious disease from affecting another."

The sailor briefly hesitated. "I have to see inside this casket."

"As you wish." The man reached into his breast pocket and produced a paper. "According to your laws, here is the death certificate."

The guardsman examined the paperwork. The pilot said his passengers had the proper paperwork.

"You must wear a mask to open the casket." A second man entered the cabin. "I am ship's doctor." *If he was the doctor, it made sense that the other man could be a nurse. Or posing as a nurse.*

Mallory saw the lieutenant don medical gloves and mask and move to open the casket. The interior must have held what the doctor said it did, or she would not be painstakingly viewing these reports, still searching for clues leading to Marc.

"I can get you a scalpel—a surgical knife to cut through the seal," the ship's doctor offered.

A second later, his Coast Guard teammate poked his head inside the room. "All clear on my side. You ready to move on?"

The lieutenant leaned forward once more, the camera moving close to the casket. Mallory heard the coast guardsman cautiously inhale. Then the picture bobbed as the man retched. He stepped back and pulled the mask from his face. "Yeah. I'm done here." He peeled the gloves from his hands and pressed them into the Chinese man's hands as he quickly exited the room.

After watching the event a second and a third time, she sat back. Mallory felt equal parts anger and hope. Furious that the inspection had been poorly conducted. That single slip may have cost her brother his freedom. Maybe his life. If Marc lay in that casket, and the possibility made sense considering the timing and the direction of the medical flight that took him west, then the Coast Guard could have rescued him. They were that close.

Seeing that the inspection had grossly missed an opportune moment, Mallory had renewed hope that she could track Marc's next destination. She enlarged the picture of the Asian man in the footage. Where did Shanghai Sumi dock? Who was this Asian man?

Chapter 16

The kibbutz moved with a quiet urgency. Men and women went about their tasks, children were tenderly tended. A palpable air of preparation mixed with the usual daily responsibilities.

Marc observed leaders in the community carry out Adi's instructions. In a short time, vehicles were readied and nearly simultaneously left the kibbutz. First a tour bus pulled out of the parking lot. Next, a van left for the Ben Gurion Airport, located nine miles southeast of Tel Aviv and named for Israel's first prime minister, David Ben Gurion. Last, a jeep drove in the direction of the previous rendezvous place where Marc and Lei had been discovered.

"We're going on the van to the airport." As the plans were executed, Marc took Lei's hand and started for the motor pool. Adi's deceptions were his affair. More like his business. Taking care of his own business, Marc would hitchhike to the next available flight west.

"No."

"It's not a question. We're going home. Now." Flying out of an international airport with regular and trusted flights to the United States ranked as the fastest way to the Midwest that Marc could think of. Crammed into a coach seat between linebackers snacking on miniature packages of pretzels would be luxurious compared to Yao's travel accommodations to China and Marc's own recent arrangements crammed into crates built to transport illegal weapons that deposited them in Israel.

Reuven blocked their path. Marc smirked at the bodyguard and realized he could no longer be intimidated. Events were in place, and Indiana appeared closer. "You're getting redundant."

From behind them, Adi spoke. "That is exactly where your hunters will look. When they find you, you will never see your home again."

Marc hesitated. He didn't want to further risk Lei. But every moment he could not devote to finding Mallory endangered his sister. He faced the young Israeli. "I have to get to the United States. One way or another, I'm going. And I'm going now."

Adi spit empty sunflower shells. "Today, we will shake your pursuers from the trail. Then you will go home. You have my word."

A short time later, Adi led Marc and Lei to the spa. Smelling of water and earthy minerals, the warm air felt heavy with humidity. The place reminded Marc of a YMCA gym, with participants busy on physical therapies from swimming to mud baths. Along one wall, thick towels hung over white paneled doors along a bank of individual dressing rooms with showers to rinse off the medicinal mud.

Balancing a tray of tall glasses filled with chilled orange and pineapple juice, a woman Marc strongly suspected was Adi's mother spoke low to Adi. "The bus was followed, and the van."

Adi reached for a glass and gave it to Lei. He took one for himself and gestured for Marc to do the same. "And the third vehicle?"

She held the tray to Marc while continuing her report. Marc took a glass. She offered drinks to three others who milled casually nearby. "The last vehicle left moments later. It also was followed."

Two men, trying hard to appear as tourists checking out the kibbutz spa facilities, peered intently at faces. A half dozen men and women lounged in the active pool, reclining against jets that massaged their feet and back. In a shallow pool, families played with young children. In the far spa, three elderly men soaked in hot mineral waters produced only in Israel.

The woman circulated, offering beverages to clients. She came to the two men and held out her tray. The taller of the two irritably brushed her off. With an oily smile, the second took a glass and nudged his companion to do the same. Her tray now empty, Esther engaged the strangers. Boldly, the woman walked them through the facility, explaining how much healthier their skin would be for a reasonable fee. One price included the therapeutic waters, massage, and medicinal mud masks.

Again, the taller man attempted to wave her away. She motioned for two therapists to join her, and the three began to sell the wonders of the spa experience in earnest. They herded the two to a counter where towels and a sheet to wrap in were placed in their hands. Talking rapidly and convincingly, one therapist guided the two to dressing rooms and pressed them inside with the promise of immediate attention from a masseuse. Suddenly panic stricken, the two pushed the bundle of towels and draping into her hands and beat a hasty retreat back the way they'd come.

Their bodies and faces completely covered with rich mud mineral masks from the Salty Sea, Adi, Marc, Lei, and two other couples, watched them go.

Chapter 17

Mallory's heart beat double time. This was the report she had waited for. When the phone rang on a secure line in her office, the caller would provide the long-anticipated update on Shanghai Sumi. Mallory was confident that the ship's destination would lead her to Marc's location.

Professionally, she knew better than to allow her emotions to run ahead of the concrete facts. Personally, Mallory fanned her smoldering hope. All she had was hope.

The voice on the other end sounded nasally with a head cold. "She docked in Shanghai."

That fit the sketchy details Mallory had gathered so far. The Asian man on the medical flight and on the cargo ship could possibly be Chinese.

Sniff. "Agents learned their military was developing advanced technology in an old temple adjacent to the port." Sniff.

Anticipation brought Mallory to her feet. The Meissner Device. The thieves were building the Meissner Device Generator from the stolen patent. Marc's invention from his patent. "Get inside that facility. Find Marc."

The caller blew his nose. "We did—"

"Did you find him? Was he there?" She gave up trying to keep her voice professionally neutral.

"There was an explosion. The project is gone."

The words felt like a sucker punch to her gut. Mallory thought about Marc's early description of the patent he submitted to the United States Patent and Trademark Office, so Mallory could track the leaks that were delivering America's best ideas to America's enemies. *The part I don't have, I really don't have.*

But the next statement was worse.

"At least four men were killed."

Her legs suddenly refused to hold her weight. Mallory slumped onto her desk chair. She rested her forehead in her palm.

"Someone meeting Marc's description had been there. A Caucasian with a ponytail."

Had Marc made a deadly miscalculation in his attempt to bridge the gap of missing science for his invention? "A man meeting that description is one of the dead?"

Sniff. "The four dead were Chinese."

She made him check his facts and repeat the previous statement three times. "Then where is he now?" This was the single question she had asked since the day Marc disappeared.

He blew his nose. "Based on what we gathered so far, I have no idea."

Mallory fought the urge to throw the phone on the floor and stomp the piece to oblivion. "Okay, there is a lot you don't know. Tell me what you do know."

"One; the Chinese military was developing something they were keeping secret in the lab. Two; someone matching Marc Wayne's description had been involved. Third; the guy we believe was Mr. Wayne is either dead, has escaped, or been taken to a new location. Fourth; if he managed to get away, he didn't leave breadcrumbs for us to follow."

Chapter 18

After a couple hours to rest, Marc and Lei were summoned. The cottage they had been given was comfortable. Marc had considered explaining that they needed separate rooms, and then changed his mind. Sleeping on the couch near Lei seemed more important than their individual privacy. He suspected she felt relieved that he stayed close.

"It's time." Drinking coffee with his cousin, Adi sat at an outside table in a garden courtyard.

"You're sending us home?"

"I'm sending you."

Lei sighed with relief and laid her head against Marc's shoulder.

"Come." Reuven filled two small cups. "Have coffee while we speak."

Now that Reuven put aside his gruffness, Marc could almost see some personality. Almost. He set steaming Turkish coffee on the table. Adi's mother brought a platter piled with bright oranges, sliced eggplant, smooth hummus topped with a swirl of olive oil, and warm flatbread. Her shoulder length black hair pulled back from her face, she set about rearranging the furniture on the patio.

Our last meal? Marc was eager to travel home. The weight of Mallory's safety felt heavier on his heart with every passing hour. He prayed she lived, that he could quickly locate her and get her away from the beasts that held her. Daily, he recalled the computer image Yao had shown when Marc refused to build the Meissner Device. Marc had watched in horror as his living room appeared on the live stream, and three men dragged a woman to the couch. As they pinned her down, Marc felt primal anger mixed with fear as he recognized Mallory's cinnamon colored hair.

"Call off your buffoons!" Marc's panic had spun out of control. "Don't let them touch her!"

"No one is going to rape her. Though it was difficult to prevent even such loyal men from enjoying the few privileges of their job." Yao had sighed theatrically. "No, she won't be raped. Not this time." He squeezed Marc's shoulder. "Though that could be arranged. Consider this mercy a trade for your cooperation."

The Colonel spoke into a cell phone while Marc watched the screen. Moving close to the camera, a fourth man held a syringe in one hand and a small, rubber-topped bottle in the other.

"Don't." Marc shuddered. "Please."

"Succinylcholine." The Colonel narrated. "An effective paralytic. Administered intramuscularly, the drug immediately depolarizes the muscles while leaving them flaccid."

From somewhere far from his home in Indiana, Marc watched helplessly as the needle pierced Mallory's arm.

Beginning at his sister's head, a rapid seizure-like movement quaked through her body. Then she stopped breathing.

"Not one muscle of her body can move." The Colonel continued in a calm, informative tone as if he described the ordinary removal of a splinter. "Her diaphragm no longer pulls necessary oxygen into her pretty lungs."

The man who had administered the deadly injection positioned a breathing bag over her face. Mercifully, air entered and exited his sister's body.

"This is me breathing for her." The Colonel whispered the words in Marc's ear. "And this is me not breathing for her." The Colonel spoke into the phone. On the screen, a wiry man with dark hair pressed a cell phone to his ear and then said something Marc could not hear to the one who held the breathing bag. The bag stopped the simple yet vital function and lay like a deflated balloon against his sister's face.

"A human typically experiences brain damage when deprived of air for longer than four minutes."

"For God's sake!" Marc had pleaded, "Give her air."

"I will provide what she needs when you give me what I require." He pointed to the time at the top right of the computer screen. "She has been without oxygen for a full minute."

Standing tall, Yao clasped his hands behind his back as if giving a college course lecture. "Hypoxia is a pathological condition in which the body as a whole is deprived of adequate oxygen. The brain in your sister's lovely head will suffer cerebral anoxia. Continued oxygen deprivation as is happening here," he pointed to the screen, "results in coma, seizures, cessation of brain stem reflexes, and," he paused to allow Marc to mentally take the next step, "brain death."

Still unemotional, the Colonel indicated the time on the screen. "It has been two full minutes."

Hungrily, Marc's eyes bore into the 12-inch screen for any hint that his sister might really be all right. That this wicked nightmare would end, and he would wake any moment. His inventive mind searched for solutions, for some way to oppose these bullies, save Mallory, and refuse the Meissner Device to these men. The device was only theoretical anyway. Even if he agreed to cooperate, it would be a hoax. If they were willing to suffocate Mallory in front of him, what were they capable of doing when they realized he couldn't make the weapon of their ambitions functional? Yet, he could not allow Mallory to die. Not because he refused to bend.

"Okay!" The word burst from Marc's lips, and he realized he had been holding his breath. "Give her air!"

"Have we reached an understanding?" The Colonel slowly drug out the words.

"Yes!" Marc slumped. "Yes, whatever you want. Just let her live."

"She is safe as long as you give me—"

"The Meissner Device." Marc filled in the rest of the sentence. "I got that part."

"Excellent." He waved toward the guard who cut Marc's duct tape bonds. "I will show you where you will assemble your invention."

That was the last time Marc had seen Mallory, and fear for his sister was a constant shadow. If Adi arranged transportation, Marc would at least be on the same continent, and hopefully in the next 24 hours. Without passports, credit cards, or other necessities of travel, Marc felt his best support would come from the American Embassy who could place a call to Mallory's office.

Marc planned to ask Mallory's coworkers with the FBI to bring him and Lei to the United States. "Have you contacted the American embassy?"

"I have made contacts," Adi assured. "But not with your Embassy."

Marc noticed the smirk on the cousin's face and wanted to smack the expression away. "What's going on?"

Under the table, Lei's trembling hand reached for him. He gave a reassuring squeeze that belied the fear rising in his throat.

Esther came to the cousin's side and said something in Hebrew. Reuven mumbled apologies and got up to help Esther move a heavy table.

"You have a value for us," Adi explained.

"You're a weapons dealer." Frustrated with his perpetual double-talk and vague statements, Marc corralled his anger before he swung a fist at Adi. "Playing all sides for the best deal for yourself."

Adi received the insult. "Things are not always as they appear, my American friend." Resting his elbows on the table, he leaned toward his guests. "You must trust me."

"Trust you for what?" Lei spoke up.

"Will you return us to the United States?" Marc held Adi's gaze. "Will you help us get home?"

Over the meaty dates and sweet persimmons, their host nodded. "As I have said."

"When?" Impatient, Marc felt that Mallory's safety pivoted on this question.

The cousin returned. "It's begun."

Marc jumped to his feet. "When? When do we go home?"

"Soon now. Arrangements have been put into place." Adi looked to Esther and Reuven. "A day. Maybe two."

Esther nodded to the two men and left.

"Two days?" Marc's hands balled into fists.

"Maybe sooner." Adi stood. "But it will not be as you think. Not as you expect."

Chapter 19

In the dim light, a slender woman with close-cropped red hair studied satellite images. The much-delayed launch of the Israeli spy satellite had finally occurred the year Shiva joined the Israeli military. With her quick eye for detail, she had been assigned to the satellite team. Fascinated with the science, Shiva memorized great quantities of information, taking to her job like the satellite took to the sky, and when her mandatory military term ended, she stayed.

Through an agreement with India, the Indian Space Research Organization's workhorse rocket, the Polar Satellite Launch Vehicle or PSLV-10, launched the TechSAR into orbit, and Israeli reconnaissance satellite equipped with a synthetic-aperture radar (SAR). After several delays due to technical difficulties, the satellite blasted spectacularly into space from the Satish Dhawan Space Center in Sriharikota. In a flawless lift-off from the First Launch Pad, the homegrown PSLV-10 carried the 300-kilogram satellite into orbit in just under twenty minutes. Nineteen minutes and 45 seconds to be exact, and Shiva was always exact.

TechSAR's synthetic aperture radar obtained clear images of small targets even in overcast conditions. All weather imaging was provided day and night. The microwaves sent from SAR penetrated thick cloud cover and dust storms, producing sharp photos, and significantly boosting Israel's intelligence gathering capabilities.

The first of its kind developed in Israel, the spy satellite primarily kept an informed eye on hostile neighbors, tracked Iran's territory, and spied on their efforts to develop nuclear arms. Circumnavigating the earth with an orbit of 450 kilometers perigee at the nearest point, and 580 apogee at the farthest point, the

TechSAR ranked among the world's most advanced space systems.

The secondary use for the satellite involved providing information to subscribers. For this unique and individualized service, the department collected significant fees that underwrote the expenses of the satellite and support personnel. Additional income funded the exploration of advancing technology.

Shiva had a list of organizations that paid for regular reports about what the eye in the sky saw regarding their particular location of interest. With a reputation for being meticulous and thorough, Shiva lived her life noting minutia. Now, studying photos of a patio, she checked the position of the furniture once, twice, three times.

Just as she had first suspected. The furniture lay positioned in the exact pattern noted in the directive. The subscriber for that seemingly trivial detail would be notified. Only the subscriber, of course, would know the meaning of the change on the ground.

Shiva would never be privy to what message that specific layout of furniture in an outdoor courtyard deep in the Negev conveyed. Nor did she believe for a moment she might be the only one consumed with particulars in the defense of her beloved Israel.

Chapter 20

At lunchtime, the restaurant at the kibbutz welcomed guests from three tour buses. Recently back from college, two young women met the groups and explained the options between the buffet and ala carte menu and pointed the way to the restrooms where travelers could wash up. As usual, the bus drivers, tour guide, and the tour coordinator received complimentary meals. The small gesture of hospitality and appreciation proved good for business.

The Japanese tourists talked fast and snapped photos faster. They moved speedily through the lunch buffet and were the first back aboard their bus and on their way once more.

The two remaining buses carried visitors from Cape Town, South Africa. Dressed in bright colored, loose fitting clothing, they sang their thanksgiving in rich harmony prior to eating.

Esther smiled and extended a welcome to the guests, glancing expectantly toward the door each time someone came in. At last, a middle-aged man in a conservative suit confidently entered and strode to the hostess.

"Shalom," Esther greeted.

"You were expecting me, of course," the man replied.

"Of course." She led him to a table near the window, her long, wavy hair swaying in rhythm to her hips.

"My favorite table," he acknowledged and took his seat.

Esther set a menu before him and disappeared into the kitchen. When she returned, she placed a basket filled with challah bread on the table. "Just out of the kitchen ovens."

"You spoil me." He smelled the sweet, golden loaf appreciatively.

"What is that between you and me? We do good business together. You bring people to our kibbutz, and we give you our best service."

"Is there also some of my favorite hummus available today? Your homemade?" A pivotal inquiry. Yes, meant she had indeed summoned him. That she had a vital message.

She smiled. "As only we make it, of course."

Once more, the beautiful Jewish woman disappeared into the kitchen. The guest watched tourists line-up for the customary buffet. His waitress quickly returned to his table with a pottery bowl of golden hummus. A sprig of fresh mint leaves rested in the shallow pool of cold-pressed olive oil.

He dipped a chunk of challah bread into the mix. "How is your family?"

"Well. The kibbutz is well." From her tray, she set before him a tall, frosty glass of orange and pineapple juice. "And congratulations on your sister's fiftieth anniversary."

He nodded cordially. "Thank you." Sister and fifty. Sister gave him reason to consider someone with a compatible relationship to Israel. Fifty indicated the fifty united states. Esther had information crucial to the United States. In this patient work, he waited on her timing and wove her words carefully to understand her message, necessarily vague to protect her and her sources.

She turned to go, then faced him again. "For such an important occasion, you should send her something special."

He patted the napkin against his lips to absorb the oil that clung there. "As always, you are correct. I should send her something special." Tucking the cloth napkin over his tie, he leaned back. "What do you suggest?"

"Such a rare occasion." Esther rested the tray on her hip and appeared thoughtful. "Some jewelry, perhaps."

"Jewelry?"

She smiled. "Diamonds."

Chapter 21

The next day, Adi invited Marc to the school's science department. Classes had ended for the day, and the classrooms were vacant except for a teacher tutoring a student through a chemistry lab. Three rows of horizontal tables faced the front of the room where a long counter with a sink served as the lectern. The setup indicated the instructor taught the course work in a hands-on fashion. Books filled shelves near the door, and the far side of the room held a variety of projects in various stages of experimentation.

When Adi and Marc arrived, shadowed as always by Reuven, the teacher encouraged the student to continue and came to meet the visitors.

Adi introduced Marc to the instructor, a man near Marc's own age with a mop of unruly curls and several days of beard growth. "Moshe is connected to the Israel Ministry of Infrastructure. He has something I think you will want to see."

Eagerly, Moshe led Marc to a circular track built of square magnets that resembled a chocolate bar. Lifting a disk from dry ice, the instructor set the disk three inches above the oval track. With a gentle push from Moshe, the disk glided above the magnets, traveling the circle in a perpetual motion at a consistent speed.

"By suspending a superconducting disk above—" he caught the circling disk and placed the object several inches below the magnets—"or below these permanent magnets, the magnetic field is locked inside the superconductor."

"Quantum trapping." Marc lightly tipped one edge down so the disk floated at an angle. Giving the object a push, as he knew would happen, the perpetual movement remained at an angle like the leaning tower of Pisa as it traveled the track.

"Exactly." Moshe pulled a pencil from behind his ear. Though he passed the writing utensil between the disk and the track, the pencil failed to interrupt or even slow the motion.

"This is your Meissner Effect," Adi said. "The reason you are here?"

Marc nodded. "The phenomenon is the expulsion of a magnetic field from a superconductor. In 1933, German physicist Walther Meissner measured the magnetic field distribution outside superconducting tin and lead samples."

The chemistry student abandoned her studies and approached to watch. Moshe beckoned her closer. "In the presence of an applied magnetic field, Meissner discovered the samples cooled below their superconducting transition temperature. At the new temperature, the tin and lead canceled nearly all magnetic fields inside."

Moshe caught the flying piece and placed the disk in Adi's hand. "This is sapphire coated with a thin layer of porcelain."

Adi turned and tapped the bottom before passing the disk to the student. With a flourish, the high schooler spun the curious object back onto the track.

She looked to Moshe. "Do you think we will travel like this in my generation?"

Excitement born of possibilities reflected in Moshe's eyes. That look reminded Marc of Hebron Heath, the enthusiastic college student who had introduced himself to Marc the last time Marc lectured at the university. Similar to when he had spoken at Lei's physics department, Marc's topic had been The Meissner Effect; Laboratory Oddity or Revolutionary Tool for Mankind. Hebron had quickly grasped the implications. The nation that developed the Meissner Effect Generator would be invincible.

The redheaded, freckled Hebron had developed a device to judge the clarity of diamonds and asked Marc what else the invention could be used for. When Marc found himself in a lab on the far side of the globe, Hebron's device was one of the few possessions his captors had brought, along with Marc, from his Midwest office. Marc had used the college student's invention to duplicate an identification badge, giving Marc and Lei access to the area where Marc had packed the two of them into crates. Marc and Lei had escaped the Chinese captors hidden in boxes used to ship illegal weapons.

Marc carried Hebron's device in his pocket. He hoped to return the invention along with the news of another use, and Marc's personal thanks.

"It is possible that we can harness the Meissner Effect." Moshe set the disk traveling in the opposite direction. "Perhaps even probable."

"What holds you back?" Marc wondered if Israel would ironically conquer the Meissner Effect before either the United States or Colonel Yao's China. Situated at the crossroads of the world, the small country of Israel had large enemies.

"This is still experimental." Moshe indicated equations on the chalkboard. "There remain aspects that have yet to be mastered."

Marc nodded. He was well aware of the unmastered elements. Or at least those aspects that continued to be elusive to him.

Adi dropped a hand on Marc's shoulder. "I have arrangements to make. Perhaps you would like to stay a while with Moshe?" Without waiting for an answer, Adi left.

Chapter 22

Hot coffee in hand, Mallory stepped into the elevator. She pressed the button for Jimmy's floor, that place where all the computer techies hung out in the digital forensics department. She leaned her forehead against the cool gray steel of the elevator wall, allowing the vibration to serve as a massage for her pounding headache. Emanating from tight muscles in her shoulders and neck, this headache had been an unwelcome companion since Marc had disappeared.

Too soon, the elevator stopped, and the doors whispered open. Deep in conversation, Deverell and Thomas waited. Thomas smiled when he saw her.

Deverell frowned. "You look like five miles of bad road."

"So do you," she grumbled, "but I wasn't going to bring it up."

He studied her. "Are you getting any sleep?"

"Based on your description, apparently not enough."

Thomas gently took Mallory's elbow. "Let's see what Jimmy has for us." He steered her toward Jimmy's workstation. Deverell fell into step behind.

Watching for them, Jimmy rocked forward on his toes. His face brightened when they arrived. "Watch this." He sat down at his dueling computers.

Deverell bent close. "You have a breakthrough?"

Once again, the picture of the cat appeared on the screen. The cat with fur embedded with a secret code. Mallory detested that image. She didn't think she even liked cats anymore.

"The only way I could jam the steganography was to turn the technology back on itself," Jimmy began. "Double stegging."

"How does it work?" Deverell wanted to know.

Jimmy swiveled his chair to address Mallory. "You told me to come up with a technique to protect places like the Patent Office. I can't figure a way to keep people from creating these files with embedded information. But I found that if we can damage at least some part of the file, the hidden encryption becomes garbled and cannot be deciphered."

Mallory rubbed her temples. "In simple terms, Jimmy."

He spun back to his computer screen, and his fingers flew over the keyboard. "Double stegging adds noise, scrambling the figure's least-significant bits. If the cat in the picture is just a cat, double stegging doesn't cause any harm."

Thomas pushed his hands into his pockets. "But for a hidden file?"

The bespectacled Jimmy nodded. "The addition of a double-stegging algorithm turns the information to gibberish."

Thomas raised his eyebrows. "Every time?"

Jimmy leaned back in his chair and laced his fingers behind his head. "An extremely high percentage of the files we tested were destroyed."

"But the process is not 100 percent effective." Mallory longed for good news somewhere in this complicated situation that had begun so simple. As the lead on her first project, Mallory's instructions from her boss were to find the leaks at the United States Patent and Trademark Office that shuttled ideas with military significance into the hands of America's enemies. "Shut them down, and while you're at it—quite frankly—get some good PR," Deverell had outlined. Now, Mallory rubbed the back of her aching neck. Instead of plugging a simple information leak, they had uncovered a problem that impacted the balance of world power.

"Like most things in life," Jimmy looked apologetic, "it's 98 percent."

"What about text files," she pressed. "Will this work on them as well?"

"I've only tested this on image file carriers," Jimmy adjusted his glasses, "but I'm confident this method can be extended to additional formats."

"Like video and audio?" Thomas handed stomach lozenges to Mallory. "Can those carry hidden messages?"

Deverell pulled a pen from his shirt pocket and began clicking. "There are so many holes for the Little Dutch Boy to plug in the information dikes."

"Relax. I've got you covered." Jimmy took the pen from Deverell, slipped it back into Deverell's pocket, and patted it into place. "Digital steganography relies on the same basic principles to conceal data for any digital carrier."

Deverell unwrapped a stick of gum. "Can you make this program available?"

"In my sleep." Jimmy waved his arm in the direction of the other geeks busily at work in their cubicles. "We'll create the software and make it available to any organization. The software will scour all outgoing communication for fiendish content."

"Won't installing the software be a heads up for anyone doing this?" Mallory spoke around the stomach lozenge from Thomas. "The last thing we want to do is give someone involved in espionage a reason to find another avenue."

"This goes at the email server level," Jimmy assured. "Every communication will be automatically filtered through an algorithm. Individual employees will not be aware of the addition of the new software."

"Unless someone tells them." After all this hard work, Mallory didn't want anyone to outsmart Jimmy's technology. Not until she got Marc home where he belonged. She wanted the tediousness of spy-prevention at the USPTO behind her, so her full attention, and the concentration of her co-workers, could be centered on locating her brother.

"Will this tell us who fits the glass slipper?"

Jimmy looked quizzically at Deverell.

Mallory translated her supervisor's question. "With this program, can you tell us who is the mole?"

"Sorry, Mallory." Jimmy shrugged. "It's easier to jam the steganography than detect its presence. I can stop the attack. But I can't tell you who is doing it."

Chapter 23

At 8:00 p.m., a tour bus parked near the entrance to the diamond exchange. From his parking place, the diplomat watched until the front and rear doors of the bus swung open. Chatting excitedly, a kaleidoscope of tourists disembarked.

The heart of world events, Israel remained a popular vacation destination for peoples of many faiths. Generations came to the cradle of civilization as a pilgrimage. From Egyptian hieroglyphics at the world's largest Roman archeological dig in Beth Shan where the Philistines hung King Saul's body, to the caves at Qumran where the Dead Sea Scrolls long lay hidden; from the aqueducts and man-made port of Caesarea Philippi to the impenetrable mountaintop fortress of Masada, the small nation offered an abundance of tourist destinations.

Two of the most popular destinations were the Church of the Nativity and the Church of the Holy Sepulcher. Marking Christ's life from the womb to the tomb by planting these churches, Constantine's mother, Helena, made her pilgrimage when she was nearly an octogenarian. While Helena's arduous trip focused on discipleship, the Middle Ages would bring a different group of travelers who came for penance. Modern day visitors came, many making consecutive journeys, for religious and intellectual reasons. Ancient history continued to be unearthed each time new construction began.

The diplomat knew diamonds were a favorite remembrance of the Holy Land for tourists to take home. Dressed in casual clothes, he left his car and mixed with the visitors. Sponsored by an interdenominational church from the United States, this group consisted of a collection of congregants on a two-week tour of the land where Jesus walked.

Inside, the group milled in a bright welcome center, reading informative displays outlining the process that transitioned the diamonds, rough from the earth, into costly precious gems. "Sixty percent of all diamonds in the world are distributed through Israel," began the tour guide. "Of those diamonds, 61 percent go to the United States, 14 percent go to China, eight percent travel to Belgium, five percent to Switzerland, two percent to Japan, and two percent to England."

The guide answered several questions from her audience. "The value of a diamond is based on the four Cs," she continued. "Color, carat, clarity, and cut."

Pointing to a colorful wall display, the cheery guide beckoned the group to move closer. "Here we receive diamonds to cut and polish. Each diamond has 58 facets. The round cut diamond represents love. The square diamond is the second most popular."

Following the flow of the building, the tour passed through a low-ceilinged hallway. Along the left side of the passageway, four small rooms were visible through generous windows. Inside each room stood a glass case. In the first of these private shopping areas, a saleswoman in high heels showed a velvet case of similarly cut diamonds to a couple seated in plush chairs.

The guide stopped in the middle of the hallway. "If any of you would like to shop for a special diamond," she indicated the adjacent rooms, "one of our diamond experts will be pleased to help you find the exact one you are looking for."

The hall opened into a glamorous showroom. Their footsteps muffled in thick turquoise carpet, the tourists walked slowly, admiring one sparkling showcase after another. Each stunning display was a collection of precious stones—blood red rubies, sea green emeralds, autumn topaz, glittery opals, and, of course, diamonds.

The diplomat mingled for several moments, admiring a case of cameos. Casually, he drifted to a case of unusual blue and green mottled stones.

"This is the Eilat Stone." The stunning dark-haired beauty at the display had a deep, melodic voice. "Named for the locality where it is found, at the southernmost city of Israel."

The diplomat met her eyes. "This gem resembles a stone common in my country, the turquoise."

"Turquoise was prized by American Indians." She set a polished stone on a soft cloth. "There is a similarity. Eilat stone is chrysocolla that is intermixed with turquoise and a form of malachite."

A small, elderly woman pulled her tall, thin husband to the case. "Isn't this one pretty?"

Her husband bent for a closer inspection. "Shades of blue and green," he mumbled in a Mid-western accent. "And matches your eyes. But I promised you a diamond for our fiftieth anniversary."

His wife took his arm, and they crossed to the next display featuring pearls.

"The Eilat Stone," the saleswoman continued, "is our national stone."

"This broach is striking." He pointed to a free-form piece. The stone looked to be the size of a half-dollar, set in gold, and ringed with tiny diamonds.

Pulling a petite gold key from her pocket, the woman unlocked the door on her side of the jewelry case. "The mines from which the Eilat came are believed to have been the copper mines of King Solomon." She laid the broach in his palm. "The gem, in fact, is often referred to as King Solomon Stone."

Gently turning the brooch this way and that, he watched the stone and the diamonds catch the light. "The diamonds are interesting."

"Natural, untreated, and responsibly sourced, these diamonds are cut and polished by a specially selected diamantaire. Called Forevermark, these diamonds have an icon and identification number inscribed on the table facet."

He peered closer at one of the clear stones that ringed the Eilat. "I don't see any marks."

"To preserve the beauty of the diamond and not detract, the inscription is 1/500 of the depth of a human hair."

"Incredible."

"Are you shopping for something particular? Does this piece please your eye?"

He held her gaze. "I'm looking for something special for my sister. She is celebrating her fiftieth anniversary."

The woman nodded slightly. "I'll wrap this for you. You'll be pleased to know it comes with a certificate."

"A certificate?"

"A certificate of authenticity. This certificate tells you everything you want to know."

Chapter 24

"Come." In the predawn hours, Adi came to the cottage door. "Your journey begins."

Marc's mind raced with thoughts about Mallory. Where would he begin to search for her? Maybe he should first contact Logan Deverell. Surely, her team had been doing what they did best when she didn't show up for work at FBI headquarters. Once Mallory was safe, Marc would help Lei track down her American boyfriend who had proposed. Could life ever again return to some semblance of the normal he enjoyed before waking up in China?

Outside, Lei appeared to imprint images of the community in her memory as Adi and Reuven led them to a long, low building on the outskirts of the kibbutz. Made of corrugated steel, fans mounted along walls circulated air. The structure resembled a metal pig barn in the Midwest, but without the stink.

They ducked into a smaller barn. Under luminous lights, a half dozen men moved among several rows of automatic weapons. In the center, balanced on sturdy tables, stood four crates. Two were sealed. Two were empty.

Lei backed toward the door and turned to bolt outside, but the cousin blocked her escape. Her shoulders began to shake, and tears spilled down her smooth cheeks. When Marc put a protective hand on her back, she turned and buried her face into his chest.

He wrapped his arms tightly around her. "I won't let anything happen to you," he whispered in her ear.

Adi brought bottles of water to Marc and Lei. "I want you hydrated." From a prescription bottle in his pocket, he produced several tablets. "Take four of these. They will make this trip far more comfortable than your previous one."

Marc nodded toward the crates. "Is this necessary? Can't we just go to the Embassy?"

Adi deposited the pills into Marc's hand. "It is necessary."

Marc surveyed his options. He wanted to run, to barrel through the men standing around them, to swing an automatic weapon like a Louisville Slugger as he had in the Chinese lab with an empty LAW. He would find his own way to the Embassy. Or the airport. Or something. Anything. They could hitchhike. Lasso a couple of camels. He pictured himself pulling Lei by the hand as they rushed the group who stood poised to prevent their escape. Adi's cousin adjusted his position, so Marc could see the gun stuck in his waistband.

Marc smirked at Reuven's attempt to intimidate. Marc knew Reuven wouldn't shoot them. And Reuven knew he wouldn't shoot them.

"Liar!" hissed Lei.

Adi again received the insult. He stepped closer, and Lei recoiled against Marc.

"Lei, you are a brave woman." Adi spoke softly. "I won't lie to you."

Marc glared defiantly. "You said we're going home." He felt like a fool. These guys owed him nothing. They were arms dealers who blatantly told him they were using him for their gain.

"I told you to trust me." Once more, Adi proffered the medication. "For your comfort." When Marc hesitated, the cousin stepped forward menacingly. "We are running out of time. Take the pills and get in the crates, or I will help you."

Marc turned, welcoming the opportunity to drive his fist into Reuven's nose. Knowing in moments he would be overtaken, the satisfaction would be worth whatever blows came his way.

Adi put a calming hand on his cousin's arm while he kept his gaze on Marc. "A man has to make his own decision."

Lei looked questioningly from Adi to Marc. Marc tightened his hold on the small woman who depended on him while his thoughts raced with the possible options. He and Lei had outsmarted some of the most brilliant minds in China to get this far. Of course, he hadn't been too bright to get into China in the first place.

Adi's steady voice broke into his thoughts. "Have you been harmed here?"

"No," Marc admitted.

"Has anything been required of you?"

Marc shook his head.

"Do you have reason to mistrust me?"

Marc looked at the crates.

Adi shrugged. "So, it doesn't look like you think it should."

Still Marc hesitated. He wanted details. Reassurance. Promises. At least for Lei. But he knew he wouldn't get that from the quiet man who stood in front of him. This young leader surrounded by loyal supporters in a country laced with multiple cultures and manifold points of friction.

"Adi," growled his cousin urgently. "We are running out of time."

Adi waved an inviting hand at the crates. "Your journey begins as soon as you are ready."

Lei looked questioningly at Marc.

"Let's go." The decision was made. Hoping he made the best choice, he pressed the pills into her hand and urged her to swallow them. He swallowed his own and escorted the tentative Lei to a crate. Unlike their previous experience, the interior of these boxes was larger and padded for comfort.

Marc pulled Lei into his arms to lift her but swayed as the medication swept through his limbs. Immediately, Adi came alongside and steadied Marc. With an assuring look, Adi lifted Lei into his own strong arms. Her eyelids were heavy, and she fluttered them in a vain attempt to remain awake. Adi looked long at the girl's pretty face before gently settling her in the open crate.

Feeling woozy, Marc barely kept his balance as he lowered himself into the second container. Aware of others around them moving and speaking, Marc noted they did not immediately close the crates.

"Are you okay, Lei?" Marc called to her.

"Yes," came the reply, though he heard the quaver in her voice.

Moments later, he jerked awake and called again. "Lei, are you all right?"

There was no response.

"Lei?" He called again.

Adi came to his side. "Your wife is sleeping now. You may allow yourself to rest."

Marc tried to shake his head. "Not my wife." He spoke each word with effort. He focused on Adi and saw confusion cross the confident man's face. The expression looked so out of sorts for Adi that Marc grinned like a drunkard, his face only half cooperating. "She's my friend," he slurred.

"But she wears your ring."

Marc couldn't resist. "Things are not always as they appear."

Adi studied his face, and Marc knew the Jewish man's quick mind considered the options. He swallowed, summoning moisture for his thickening tongue. "Not my ring," were the only words he could speak through lips that refused to cooperate.

Marc's vision clouded, but before he permitted himself to sink into the seductive darkness of sleep, he forced his eyes to open once more. Rising slightly, Marc saw Adi bent over Lei's crate. The tall Israeli lightly kissed the sleeping girl on the forehead. Or maybe Marc was already dreaming.

Movement prodded Marc awake. The simple act of opening his eyes required a herculean effort. He had one last vision of Adi, slipping a small package into Marc's pocket. From somewhere far away he heard, "Yala yala!"

Chapter 25

"Holy Buckets! Hey, Major, look at this!"

A thunderous crack and loud voices woke Marc. He opened his eyes and shaded his face from the sudden brash light.

"Our bosses back home ain't gonna believe this," came another voice. "Check out those other crates."

As he heard the tops wrenched from the other boxes, Marc's vision began to adjust. He struggled to orient himself. Where was he? Who were these guys? Feeling hungover, he grabbed the side of the crate and hoisted himself into a sitting position.

"Whoa, easy there partner." The voice carried a distinctly southern accent. Marc blinked and squinted at the man who stood over him. His heart jumped when he recognized the American military uniform.

"American?"

"Yes, sir. Sergeant Cooper, sir."

His voice choked with emotion, Marc suppressed an urge to hug the soldier. Instead, he stretched out his hand. "You have no idea how glad I am to see you."

The soldier clasped his hand in a bone-breaking handshake. "Yes, sir."

"There's a girl …" Marc began groggily. From somewhere nearby a voice exclaimed, "Major, there's a girl in this one."

"No way." Cooper trotted away from Marc.

Stretching the kinks in his cramped muscles, Marc stiffly crawled over the side and out of his coffin-like bed. Israeli soldiers pried open similar boxes and examined the weapons inside. Four men who looked a lot like the first guys Marc had met in Israel were on their knees, hands clasped behind their heads. An Israeli stood over them with a beretta.

As a second Israeli secured the wrists of the prisoners with plastic handcuffs, a kneeling man twisted and grabbed the throat of his captor. Before the other prisoners could react, the Israeli broke his attacker's hold and crushed the Arab's windpipe. For the next four minutes, the three prisoners watched their companion writhe and struggle for oxygen. At last, the man's eyes glazed over, and his form went lax. The horrifying sight of his suffering proved a deterrent to any further aggression.

Marc stumbled over to Lei's container where three men in American uniforms were gathered.

Cooper glanced toward Marc and back to Lei. "That doesn't fit with any intel."

"We knew we would find weapons. And a special package people in suits have been looking for." The Major pierced Marc with his gaze. "But nothing about a second special package."

Marc pushed his way to Lei's side, relieved that he had arrived before she woke and panicked. "Lei." Gently he shook her. "Wake up. The cavalry arrived."

Chapter 26

With astonishing efficiency, the four desert rats—one in a body bag—were loaded at gunpoint into an Israeli military truck. The four Israeli military men, including the one who broke the neck of his prisoner with the ease of answering his phone—prepared to leave.

Marc knew they must be spec ops. One spoke briefly with the American Major. The American and the Israeli soldiers shook hands, and the Israelis, with their prisoners and shipments of arms, were gone.

His arm around Lei's shoulders as an assurance for her, Marc took in the scenery. They were still in the desert though in a much different place than they had known with Adi.

Marc looked down at Lei. "He did it."

Lei smiled wistfully. "Adi got us to the Americans."

An airman jogged to the helicopter that rested on the ground like an oversized insect. Marc could see the pilot begin the sequence for take-off.

"Major, we can't take the girl." Cooper lowered his voice, but Marc heard him.

Motioning for Lei to remain where she was, Marc moved closer. "She comes with me."

The Major shouldered his weapon. "At ease, Sergeant. Major Northington once brought home four girls who weren't part of the mission."

Cooper wasn't easily swayed. "Begging your pardon, sir. I thought that was a legend."

"Legendary."

"Our orders, sir..." Cooper reminded through clenched teeth.

His nose near Cooper's, the Major was authoritative. "I'm well aware of our orders."

"We're still gonna get our butts chewed."

"That's customary." The Major motioned for the group to move out. "The girl goes with us. Now, everyone in the heli."

Chapter 27

Mallory found Deverell already in the conference room when she entered. Thomas followed close behind. Frustrated to have her investigative nose at a dead end, she voiced as much to her boss. He agreed to hold another pow-wow.

Papers surrounded his laptop, a map lay to his left, and reading glasses perched low on Deverell's nose.

"What's all this?" Thomas picked up the map.

"More surface space." Deverell mumbled without bothering to look up.

Mallory combed through the top layer of papers. "Anything in here that tells us something about Marc?" Having trailed her brother to Shanghai, she hit an impasse. She didn't know where to search next. Had he been hurt in the explosion at the lab? Carted off to another high security facility? If Marc had gotten away, why had he not made contact?

Deverell took back his papers and patted them into their previous pile. "You know I'll tell you as soon as I have anything new." He removed his glasses. "You two sit down and cool your jets. The purpose of this meeting is an exchange of information. To catch each of us up on what we have to date."

Thomas sat. She paced.

"Mallory," Deverell said wearily. "Will you please sit?"

"No."

"Your incessant pacing is distracting to say the least."

"So is your incessant pen clicking." She eyed him fiercely. "And gum popping."

"Touché," Thomas pronounced. "Now—"

The door flew open, and Jimmy bounced inside. Wearing a Hawaiian shirt and Birkenstocks, he looked more like a hippy than a computer genius. “Here it is.” He spun a flash drive between the pointer fingers on each hand.

“Here what is?” Deverell indicated a chair.

“What you asked for.” Jimmy spoke as if reminding a forgetful child. “The double-stegging software, though the guys in computer forensics call it steggo-stomping, they even put a dance to it—”

“Jimmy.” Mallory snapped her fingers. “Focus.”

He froze, looked again at the flash drive in his hand, and started fresh. “This server-level technology will filter outgoing email.”

“This will mitigate major espionage?” With his foot, Thomas pushed out the chair Jimmy had so far ignored and waved Jimmy into it.

“The problem in the past,” Jimmy dropped into the seat, “is that as soon as security personnel figured out how to circumvent one algorithm, the spies easily developed ten more. This,” he waved the drive, “is double-stegging. It provides a stopgap.”

Thomas leaned forward. “How effective is your program?”

“No matter how sophisticated steganography methods become, those technological advances could be used against them.” Jimmy pushed his John Lennon glasses up on his nose.

Deverell frowned in confusion. “Them?”

“Malefactors,” Jimmy clarified.

“You mean the big bad wolves.”

“How does it work?” Thomas pointed at the drive.

“By attacking the applications using the applications themselves, the algorithms become their own worst enemy.”

“The wolf huffed and puffed but could not blow the brick house down.” Deverell held out his hand, and Jimmy dropped the flash drive into his outstretched palm.

“What about tracking down the mole?” Mallory stopped pacing and planted her hands on her hips.

Jimmy shook his head. “Sorry, Mal. While this application runs quietly on the email server, that allows time for an agent to seek out the intruder while remaining confident that no out-going mail is exporting hidden files.”

Thomas frowned. “What does that mean?”

“Thieves use this technique to make uses of static carriers like JPEG or MP3 files.” Jimmy held out his left hand. “That’s the bad side. The good side,” he held out his right hand like the second tray of a weight scale, “is that steganography is a moving target. Now, exfiltrators are making use of streaming data technologies like VoIP—"

Deverell waved him to a stop. “English. We speak English here.”

“Voice over Internet Protocol.”

Thomas raised his eyebrows. “You mean phones?”

Jimmy nodded. “That’s what I said.”

“Right,” Thomas conceded.

“As I was saying,” Jimmy eyed each of them over his glasses to be certain they were listening. “Disrupting or detecting hidden transmissions inside real-time phone calls is the next challenge for digital forensics. And considerably more complicated.”

“Job security,” Thomas noted.

The door burst open for a second time, and Deverell’s assistant rushed to his side where she bent and said something close to his ear. His eyes grew wide, and he looked immediately to Mallory.

A second later the assistant left again, the door closing smartly behind her.

“What?” Mallory held her breath, feeling hope beat an anxious rhythm in her pulse.

“We found him.” Deverell grinned. “We found Marc.”

Chapter 28

Showered and freshly clothed, thanks to the U.S. military hospitality, Marc and Lei were ushered aboard a military transport.

"It's not exactly the airlines so don't expect pretty stewardesses with pretzels and peanuts," the Major quipped. "But you don't exactly have passports."

Marc glanced appreciatively at the bare interior. "For the record, I didn't exactly have a passport to get to China."

"Israel?"

Marc shook his head. "No stamps in my passport for any of these international stops."

The Major shouldered his way to where they would sit. "I heard that about your initial trip east. Well, this may not be first-class seating, but it's a step or two above the transportation mode you were using when we found you."

Buckled into seats, the passengers consisted of Marc and Lei, and the men who had uncrated them from the journey Adi set into motion. Lei finally relaxed after the big plane settled into a westward course, and she dropped into a deep sleep. While most of the passengers slept, Marc chafed. Whether excitement to be headed home, residual adrenalin, or exultant relief that his escape plan had come together even with the unexpected added, he couldn't be sure. Maybe a combination of all of those. Plus, a healthy serving of anxiety about Mallory.

I'm going home. After the seemingly endless days of frustration, what he had most desired was finally reality. Marc took in the military plane, the rough yet trained men who had abandoned their own safe environments stateside to rescue him and Lei in dicey surroundings. Marc allowed a tidal wave of gratitude wash away the persistent fear.

Closing his eyes, he imagined his home that smelled of cherry pipe tobacco and recalled driveway basketball with Mallory. Weekdays were spent riding his bicycle to his aged brick office where he heard Dr. Thurmond's noisy four-legged patients in the adjacent veterinary office. And inside his workspace, Marc worked alongside his efficient and charming receptionist, Violet, who frequently left him tongue-tied.

After their escape, surely Mallory no longer served as leverage for Yao to make Marc do his bidding. He dared to hope Yao ordered his bullies to leave his sister sleeping on the living room couch. Marc grew tentatively optimistic about his sister's well-being.

"How's Violet?" Mallory often asked the same question when she telephoned. Always with that tone in her voice that hinted at something he should notice. He wanted to hear Mallory ask again, and he wanted to give a knowledgeable reply because he wanted to really know about Violet. What did she like about the music she played at her desk? What did she do when she wasn't working? Why did a bright and talented woman work in his one-man office when she could easily have a thousand better careers?

Picturing Violet's blue eyes, he fell at last into a deep sleep.

Later, turbulence shook Marc awake. Feeling the plane descend below the rough air, Marc swallowed to release the pressure in his ears. As quickly as the rough air had come, the disturbance ended. Shifting positions, he sought sleep once again. When the return to slumber eluded him, he left his seat to stretch. Walking felt good. The others dozed with one exception. He stopped beside the Major who was reading.

"Good book?"

"Therapy." The Major folded down the corner of a page to hold his place. "Takes me longer than the others to relax after a mission."

"A perk of being in charge?"

"You should know. You just commanded your own mission." He gestured to an empty seat.

"I gladly leave that to you trained professionals." Marc shook his head. "Those Israeli guys with you were intense."

Removing his reading glasses, the Major cleaned them on his shirt and peered at Marc through the lenses, much like Dr. Thurmond did. "The Mossad are probably the world's most sophisticated, law-approved killers."

"Israeli special forces?"

"Skilled in poisons, long and short-blade knives, and explosives the size of a cough drop that can blow off a man's head. Use an arsenal of guns from short-barrel pistols to sniper rifles with a milelong killing range. Knew a guy who used piano wire to strangle." He tucked his glasses into his pocket. "They can take out a target without leaving a mark."

"They wanted the arms dealers."

"The Mossad invited us along."

Marc digested this information. Adi's connections were at his nation's highest level. Adi had delivered them to the contacts that were certain to get Marc and Lei safely home. "I would be interested in the story about your Commander. The guy you mentioned when we met. Northington, was it? It seems to have made a difference about bringing Lei back with us."

The Major rubbed the stubble on his chin. "I could trade you that yarn for your story about how you got yourself packed into an illegal arms box like a canned Vienna sausage."

Nearby, Marc noticed Cooper stir and lean their way while working hard to appear that he wasn't listening. Marc sat. "Deal."

"I'll bet an MRE that your tale beats this fiction the pilot loaned me." The Major closed his book.

Marc recognized the author's name as a popular Korean War veteran turned novelist who crafted bestselling adventure suspense novels. He also acknowledged that this book probably released while Marc had been holed up as the Colonel's unappreciative guest. He wondered what else had happened while he was out of touch. While a captive of the Colonel's, the few conversations Marc had shared with fellow scientists in the Shanghai lab had shocked him. The People's Republic strictly censored the information provided to their citizens. He guessed that even this educated segment was about two decades behind the modern world in their awareness of global events and technology.

The Major began his story first. "In the late 80's, Major Michael Northington was assigned to bring home a senator's daughter. He found her in Asia, but she wouldn't leave without a bevy of orphan girls."

"She wanted to adopt them?" Marc knew his own adoption was not international, though many people his parents had socialized with had adopted their children from other nations.

"She wanted to keep them alive. They were on their way to having their organs harvested for the black market."

Cooper leaned in, no longer pretending that he wasn't listening.

"I'd heard rumors, but …" Marc swallowed. "I guess you guys see all kinds of nasty business." During law school, Marc had learned the basics of many types of law including criminal law. He quickly realized he didn't have the stomach for dealing with people's vile treatment of others, nor the desire to engage in a battle of wits over degrees of evilness and the rights of those who preyed on others. He neither thrived on nor found such drama appealing. That brief taste became the reason he specialized in intellectual property law. And his specialty came in handy for his own tinkering, as Violet referred to his inventions.

"There's no lack of bad guys in the world." The Major crossed his arms over a muscled chest. "Northington obeyed orders. Strictly. Like a good soldier is trained to do. Right up until he saw the remains of one of the girls."

Marc grimaced at the idea that someone would take the life of a young girl as if she were nothing more than a car to be parted out. "How many girls did he bring home?"

"Four. Plus the senator's daughter, of course. But he always regretted not making his decision before the fifth was killed. That experience still haunts him."

"And we're still gonna get our rear ends chewed." Cooper rubbed a hand over his military issue haircut that was due for a buzz. "Geez, that guy has some colorful insults, and he never repeats any."

Marc looked puzzled.

"The former Major Michael Northington," the Major explained, "is now our commanding officer."

Chapter 29

"China?" Mallory faced Deverell. "Or Israel?"

After the hours, days, and weeks of panicked suspense, she wanted details before she allowed herself to celebrate. What was Marc's condition? How much had her decisions harmed and affected her brother? She knew the repercussions would be long term. Right now, she needed to know what damage control was required immediately.

"This little piggy went to market, this little piggy stayed home." Deverell quoted on his way out the door. "Both, actually."

"That makes sense." Thomas trailed the other two.

Hard on the heels of her retreating boss, Mallory had a thousand more questions. "Where is he? Is Marc all right?"

Thomas tagged behind, shadowing the fast-walking supervisor and the power-walking analyst demanding information. "Those two countries began trade relations long before the actual establishment of diplomatic relations in 1992. No doubt conservative, the current figures indicate the bilateral trade between them is in excess of five billion."

Mallory waved an arm behind her to silence Thomas. "Will you shut up! I can't hear Deverell." Rounding the corner into his office, Deverell picked up his phone. Behind him, Mallory and Thomas followed like ducklings. "Is he safe?"

Deverell motioned for her to back away and be quiet.

Pressing near to overhear his phone conversation, she ignored his gestures. "Well? Where is Marc?"

Deverell snapped his fingers to get Thomas's attention. He pointed to Mallory and to the door. With a slight nod, Thomas took Mallory's elbow and guided her back into the hall.

"I want to know what is going on!" She jerked her arm free.

Quickly, Thomas took her arm once again and steered her out of the office, closing the door tightly behind them. "Let him do his job, Mal. You'll know soon enough." Positioning himself between his anxious partner and the door, Thomas stood sentry while he continued connecting information. "The Israeli Ministry of Industry and Trade defined China as an export target country and went fishing. China was an easy catch."

Mallory was only half listening to his constant blathering. "What are their primary fields of trade?" Mallory placed her hands on her hips. "Stolen U.S. patents?"

"Israel's main exports to the Chinese are telecommunication, high tech, agro-technology, security, and environment and info-structures."

Mallory began to pace. "If anyone hurt him …"

Thomas paced right to left, passing Mallory who paced left to right. "The Israeli Minister of Industry, Trade, and Labor made an official visit in 2008 to promote trade ties and cooperation between the two countries."

"Apparently, the two are quite chummy." Mallory faced her partner. "Where is Marc?"

"It's more than that, Mallory." Thomas spoke gently. "Israel is exposing local Chinese industries to Israeli companies in their fields. Each year, business delegations are conducted to various provinces such as Guangdong, Sichuan, Yannan, Hainan, and Heilongjiang. A trade representative office is established in southern China's city of Shenzhen and in the northeast city of Dalain."

The door opened, and Deverell beckoned them into his office. "To market, to market to buy a fat pig." Deverell parroted in singsong. "Kosher, of course."

"China imports from Israel and exports to them as well." Mallory waved for her boss to hurry with his explanation. "I know the drill."

"Home again, home again, jiggity-jig. Israel also knows that China arms Israel's enemies. Trade agreements allow Israel to have a better finger on the pulse of what is being supplied. And to who."

Mallory pointed to Deverell and then to Thomas. "What does any of this have to do with Marc's current location? What are you two hiding from me?"

Deverell stuck a piece of gum in his mouth. "It is through one of these situations that we were able to track Marc."

"You tracked him?" Mallory's voice rose an octave. "You tracked him. You knew something and didn't tell me." She drew herself up and pushed her face close to his. "How dare you not tell me—" From his desk, she grabbed a stained Harley Davidson coffee mug and flung it at the wall. The cup flew into pieces, and large chunks of glass dropped onto the carpet.

Deverell put up his palms. "Bad choice of words. Sorry. And in my defense, this information came through fast. It was through one of those channels," he held up the phone receiver, "that we were notified of Marc's location."

Through gritted teeth, she repeated her question. "How is he?"

Deverell shook his head. "I don't know."

Chapter 30

Marc felt the plane begin the descent. Gazing out the window, he recognized farm fields stretched as far as he could see like the patchwork quilts his mother used to make with the other church ladies at their weekly quilting bee. There was one on his bed back home. The women had pieced the colorful fabric squares with tiny stitches as a gift when he became a patent attorney.

Creeks and stands of trees edged square farm fields where farmers and their sons hunted deer in November. Flanking traditional red barns, newer pole barns housed oversized green and yellow farming equipment, beef cattle, and a variety of 4-H projects the country kids would display in the summer at the state fair.

A flash caught his eye. Like lightning, three F-16s streaked past, white exhaust marking their trail. The sight of them brought a lump to Marc's throat.

Then, coming into view below, was an expansive military base. Huge hangers, box-shaped dorms, and oversized buildings adjacent to the runways.

"Look familiar?"

Marc looked up to see the Major leaning over to look out the window. "Definitely Midwest." Marc craned his neck to survey the size of the runways. "Wright-Patterson Air Base?"

"Yep."

Marc called Lei to the window.

Her eyes grew large as the view filled the window. "It looks like a city."

"That base is the size of a medium community," the Major agreed. "Has everything a city has from shopping to a movie theater, and childcare centers. At last count, some 25,000 military and civilian employees work there. Wright-Patt is the largest employer in Ohio, and one of the largest Air Force bases in the world. The current economic impact to the Dayton, Ohio, region is $4.4 billion."

"Billion with a B?" Marc whistled.

"Your tax dollars at work."

"Trust me, I won't complain about taxes ever again." Suddenly, Marc even looked forward to filing the previously grievous forms. If that was how these guys got paid, he wanted to add a significant tip.

"Wright-Patt has a rich aviation heritage," the Major continued. "In 1904 and '05, on Huffman Prairie Flying Field, the Wright Brothers flew the first turn, circle, and figure eight. From 1910 to 1916, the brothers operated a flying school. Among their 119 students was Henry 'Hap' Arnold."

"The guy who commanded the Army Air Forces in World War Two?"

The Major nodded. "And H. Roy Brown, the Canadian ace."

"He shot down the Red Baron in World War I." Cooper stretched as he joined the group by the window.

"Just off the end of Runway 23 is a replica of the Wright Brothers 1905 hanger and catapult launcher." The Major pointed to the site. "That's the exact location of their early aviation accomplishments."

The sound of the engines increased an octave as the plane aligned itself over the tarmac. With a swift drop and a bounce, they were back on terra firma. Cooper swore. "Where'd that pilot get his license?"

As the soldiers shouldered their gear and made their way toward the door, Lei came to Marc's side. He nodded toward the exit. "Welcome back to the United States."

"Truly?"

"Yes, ma'am," the Major assured. "Smack dab in the center of the country."

Tears filled her eyes, and Marc put a comforting arm around her shoulders. "You can relax, Lei. You're safe."

The Major addressed Marc. "What about you? How are you feeling about landing back in the U S of A?"

Marc took a deep breath and considered. "Strangely, I feel nervous about meeting people I already know."

"Re-entry." Cooper shook his head.

"There's a term for this feeling?"

"That's the technical term." The Major jerked a thumb in Cooper's direction. "Cooper has a more descriptive slang word for it if you're interested."

Chapter 31

Their rescuers stood aside to allow the Major to lead Marc and Lei off the plane first. The aircraft's pilot and co-pilot stepped from the cockpit in time to say their good-byes.

Behind him, Marc heard Cooper's drawl. "Did we land or were we shot down?"

The pilot shot back, "You wanna try it?"

"A monkey could do it better."

"Then you're certainly qualified," the pilot responded.

Stepping into the Ohio afternoon, the first thing Marc saw was Mallory, bouncing on tiptoes and waving a cellophane wrapped bouquet of flowers. *Mallory!* His knees buckled under the wave of shock and relief. His last sight of her on the computer screen had given him nightmares ever since. Mallory! She was safe.

And here.

All his anxiety about locating her and getting her away from Yao's hired brutes suddenly dissolved. The intense shift felt like an emotional whiplash.

She ran straight into his familiar hug, the rustle of tissue and smell of flowers all part of their embrace. "You're okay." His voice choked. "You're all right." He repeated the words over and over, drinking in the reality of his sister's unexpected presence.

She held him tightly, and he must have held her even tighter because after a long moment, she squirmed and exclaimed, "Hey bro, sissy can't breathe."

He pulled back and looked at her. "That's why your eyes are wet."

Quickly, she brushed her hand across her eyes. "Allergic to the flowers." She pushed them at him.

He blinked back powerful feelings of relief and joy at seeing her. "Me, too." He held her shoulders and studied her face. "Are you hurt? Did those fiendish ogres hurt you?" He shook his head at the stupidity of the question. "Of course, you were hurt. And traumatized. I saw what they did to you—"

"Marc," she put a palm against his cheek. He saw worry and fear in her eyes. "Marc, I'm fine. Aside from horrible worry about you, I've always been fine."

A shudder went through him. "But they had you, I saw them inject …"

Stubbornly swallowing back tears, she shook her head. Gently and slowly, like she was speaking to a confused child, she chose each word carefully. "No one had me. No one injected me with anything. I've always been right here, either at home or at work. I've always been looking for you."

Marc searched her eyes. She told the truth. Now he wrestled to exchange what he thought had been reality. To trade out what Colonel Yao had shown on the computer screen with the new information Mallory just provided. He dropped his hands from her shoulders.

Mallory seemed to sense his turmoil. "Breathe," she coached. "Just breathe."

Obediently, he sucked in a lungful of clean Midwest air. Like one of his physics equations, he sorted through the sequence of events, reconciling this puzzle. Mallory had nothing to gain by lying. On the other hand, Yao had everything to lose if he couldn't make Marc believe Mallory was in danger. Once Marc had convinced the Colonel he would die before giving in to his captor's demands, Mallory was the only motivation that ensured Marc would complete the Meissner Device. His hatred of the man seethed anew, and he cursed. Smoke and mirrors. Somehow the Colonel had staged a scene that convinced Marc that Mallory's life hung at stake, dependent solely on Marc's cooperation and contribution.

"Marc?" Mallory put a hand on his arm.

He met her concerned gaze and grinned, the movement stretching facial muscles that had not been used in a long time. When was the last time that he had smiled because he felt happy? "That's the best news, sis. The very best news."

Relief showed in her eyes.

Marc put the flowers in Lei's arms and pulled her forward. "Lei, this is my sister. I suspect she had something to do with our meeting in China."

Mallory shook Lei's hand and cast an apologetic glance at her brother. "Guilty as charged. Marc has a lifetime of leverage over family vacations." She linked arms with Marc and Lei and steered them toward a waiting car.

"Wait," Marc told her. He turned back to see the Major and his men, heavy 72-hour packs thrown over their shoulders, stride to the terminal. "Major!"

The small squad halted, and Marc jogged over to them. "Major," he began. "Thank you ..." He flushed at his frustrating inability to find the right words. To express deep gratitude. These guys had risked everything and given Marc back his life.

"Next time you decide to run your own spec ops mission, call me. I'll be your support." The Major saluted.

Marc stood speechless, and the Major turned and walked away. The others nodded at Marc and followed their Major. Cooper stood a second longer to give Marc a thumbs up, then he too was gone.

Chapter 32

Marc and Lei were taken straight to Wright-Patterson's 88th Wing Medical center.

"Maybe a little worse for wear." Marc tipped his head to each side to stretch neck muscles. "But I'm sure nothing a few days of great food and some rest won't cure."

"You don't have a say." Mallory felt nearly exuberant with the quick change of events that placed her back in her comfortable position of being in control. "The military will check you thoroughly and let you know how you are."

Lei looked nervous, and Mallory patted her arm. "Don't worry. We arranged for a kindly female doctor for you. You've had plenty of excitement, and this lady is like a grandmother to all of her patients."

The hospital came into view through the military vehicle's front window. "You're in good hands." Mallory knew she chatted nonstop, but tossing information was a means of connecting to Marc. Their shared common language. And therapy for her nervous happiness. Marc had come home. Her first project as lead agent hadn't ended as a complete disaster. The most important factor of all, her brother returned safe. "This Air Force medical facility is the third largest and handles 300,000 outpatient visits annually. From here, they deploy trained medical personnel to support global exercises and operations."

"Is that all?" Marc whistled appreciatively. "This whole base is impressive."

"This wing," she added, "oversees the largest service division in the U.S. by providing fitness, food, library, recreation, and youth programs for past and present base personnel and their families."

"What do they do for fun?"

"The U.S. Air Force Marathon." Today, Mallory felt she could run and win a triathlon. "They are famous for hosting the annual U.S. Air Force Marathon each September, attracting thousands of the nation's top long-distance runners."

"Darn," Marc lamented. "Missed it by that much."

"Maybe next year," Mallory quipped.

The driver stopped at the hospital entrance.

"Only if you run it with me," he returned.

She grinned. "I'd rather beat you at driveway basketball."

Chapter 33

As Mallory promised, a grandmotherly physician took Lei under her care and gave the girl her full attention. Marc winked an assurance to Lei and turned to see what Mallory had lined up for him. A large athletic man who looked like he just came off the playing field from sacking a Colts' quarterback stuck out a meaty hand and introduced himself to Marc as Dr. Williams.

"How come she gets the nice, knit-you-a-sweater doctor, and I get John Henry who can out-hammer a machine," Marc complained quietly to his sister.

Mallory clapped her brother on the back. "Just do what the good doctor says, and you won't get hurt."

His voice still low, Marc pressed the point. "Do you cage him after hours?"

"I get extra days out when they need me to help Air Force beat Army." Dr. Williams folded his arms across his beefy chest. "I disassemble 'em on the field and bring 'em back in here to put the poor suckers back together again. It's a package deal."

"Sorry." Marc gave Mallory a "didn't know he could hear me" look.

Williams cleared his throat. "You comin' or what?"

"I'd rather 'what,' actually." Casting a last sheepish glance to his sister, Marc entered the exam room Williams indicated.

In the course of a few hours, Marc was prodded and poked in places he didn't previously know he had. Blood tests were run including a full body x-ray. At last, lying in a hospital bed enjoying the stiff, clean sheets, he smiled when Mallory entered with two tall cups smelling strongly of rich coffee.

She eyed the IV bag plugged into his left arm. "Do you want to drink this or just have it fed intravenously?"

"Don't deprive my deprived taste buds of American coffee," he begged.

She handed him a cup. "Made the military way."

"How's that?" He held the cup under his nose and breathed deeply, the fragrant steam warming his face.

"They put a bullet in the pot, and when it dissolves, the coffee is ready."

"Just the way I like it—bulletproof." He took a sip and closed his eyes, tasting the rich liquid.

"What are you thinking?"

He opened his eyes and saw Mallory's concerned look. He waved her into the nearby chair. The vinyl cushion squeaked as she sat.

"Just remembering the taste of Turkish coffee. It has a spicy, sweet flavor and grounds at the bottom that you either have to be careful to leave or chew when they swish around your teeth." He swirled the coffee and watched the miniature vortex in the cup. "It's a different color, too, sis. More orange."

"How are you?"

He looked up at her. "Nervous about a lot of things. It's a strange feeling. I wish I could blend in, but I feel like I'm being noticed. Stared at."

"We see this a lot, Marc. It's a normal part of adjustment."

"Not permanent?"

"I promise. And gives you an appreciation for what our soldiers—our military—experience with reentry."

"Those guys who brought me home?"

"It's their job. And they go through a debrief and reentry. It is hard on marriages. The ones who seem to do the best spend the first few days in a neutral place with loved ones before going to their civilian address."

Marc downed the last of his coffee. "What did Dr. Grandma say about Lei?"

His sister nodded toward the nearly empty IV bag. "Much like you, she's getting a large bag of fluids and supplements. Overall, she's in good shape considering her adventures, which you can fill me in on tonight." She cocked her head. "Now spill. Who is this pretty girl you brought all the way home from China? My future sister-in-law?"

Marc's eyebrows shot up. "Whoa. Down girl. I just got back into the states, and you're marrying me off. One shock at a time, please." Last time they were together, Mallory had been talking about Violet.

She crossed her arms. "Got it. You need time to date."

"It's not like that." He adjusted the hospital bed, so he was eye level with his interrogator. "We're not dating. She's," he fished for the right words, "she's like a student. Well, she was a student, but graduated now." He fumbled forward. "Lei is a friend who needed to get out, too, so we got out together."

"She likes you."

Annoyed, Marc shifted. "That's what Adi thought, too. Escape from a barbaric situation, and suddenly, everyone links your names together."

"I'm a research and analysis specialist. I just call 'em as I see 'em."

"She is brave and terrified at the same time. I tried to protect her, though I doubt I did a very good job. I'm weak on my superhero skills."

"You did something right. You're here."

He recalled the first time he spoke with Lei. How she spun the ring on her left ring finger. "She is in love, actually. Lei took the risk to return and find her boyfriend. She's engaged to a boy from college."

Mallory considered this. "Okay. I wondered if you were as clueless about Lei as you've been about Violet."

He frowned. "Violet? What does Violet have to do with this? She's all right, isn't she? And the office?" Fresh panic washed over him. Had Colonel Yao and his goons threatened Violet?

His sister rolled her eyes. "Violet and your office are fine."

He sighed in relief. "So, what about Lei? Medically?"

"The doctor said she is shell-shocked and exhausted. Post-traumatic stress syndrome. She's prescribing several weeks of rest, refreshment, and excellent nutrition."

"I hope I get the same prescription."

"Doubtless, Dr. Williams will offer something similar, but I'm ordering R and R and some great meals."

Marc settled back on his pillow. "Ah, steak, pizza, brats and sauerkraut, Amish-made pies, and good ole' American burgers. Lots of 'em."

"You'll die of scurvy if you don't get something green on that menu."

"Of course, what was I thinking? Plenty of mint ice cream."

Chapter 34

Released by the medical facility doctors the following day, Marc and Lei were escorted into a waiting Humvee. Much to Marc's relief, Lei looked much perkier. Mallory was safe and well. Marc told himself he could relax, but the doctor had warned him the process wasn't always immediate. What Cooper and the Major referred to as reentry.

"Where are we going?" Feeling better, Lei talked more today.

"There's something we want to show the two of you." Mallory sounded mysterious.

Marc and Lei exchanged glances. Marc wondered if he had developed an aversion to surprises. After several turns, the driver stopped in front of a building. Lei tensed at the sight of the armed guards.

"We're home, remember?" Marc looped his arm about her shoulders and gave her arm a squeeze. "These are the good guys." He smiled and held her gaze until she relaxed and smiled back.

Through several checkpoints, Mallory ushered them inside where they were joined by two others.

"Thomas!" Marc exclaimed. "What are you doing so far from your D.C. office?"

"They let me out for special occasions." Heartily, Mallory's co-worker pumped Marc's hand.

A tall man with a tanned face and glasses stepped forward. "This is my boss," Mallory introduced. "Logan Deverell."

"So, this is your little brother." The man clasped Marc's hand. "Well done. You checked out worn and weary, but healthy."

Marc looked about. "What is this place?"

"It's our lab for the development of the Meissner Effect Vehicle," Deverell announced proudly.

Lei turned and started for the door.

"Hey," Deverell called. He looked at Marc. "Was it something I said?"

"As a matter of fact." Marc jogged after Lei. Catching up, he walked in step beside her. "Could be fun to compare," he offered. "Strictly on a scientific basis, of course."

Mallory came up on Lei's other side. "You okay?"

Lei stopped and put her face in her hands. Marc drew her into his arms and tenderly cradled her against his chest. "Trust me?" He asked into her hair.

She nodded.

"Just breathe," he coached.

In a moment, she took her hands from her face and slid them around his waist. Mallory's eyes met his questioningly. Marc winked his okay.

"These people are our friends. Maybe not smart," he smiled over Lei's head as Mallory wrinkled her nose at his jab, "but you are free here to make your own choices. You are free to come and go, to choose what you want." He tightened his arms around her. "I think they just want to show us something."

Lei pulled back to study his face. Marc brushed a couple tears from her cheek. "Wanna take a peek? I bet they don't let just anybody in here."

Mallory stepped closer. "Only VIPs, Lei. And that's what you are. A very important person."

Lei nodded, and the three made their way back to Deverell and Thomas.

"Right this way." Thomas pushed the confused Deverell into the lead. "We settled on this location for a variety of reasons. Wright-Patt is one of the largest, most diverse, and organizationally complex Air Force installations. From the pioneering flights of the Wright brothers to the development of today's most advanced aircraft and aerial systems, it's all here.

Missions for the base's units vary from research and development to advanced education, and flight operations."

"This base," Mallory added, "is headquarters for a vast, worldwide logistics system, a world-class laboratory research facility, and is the Air Force's foremost acquisition and development center. There are some 60 associate organizations contracted with the Department of Defense activities."

Deverell took his cue and jumped easily into the conversation that kept pace with their footsteps. "Wright-Patterson Airbase is the birthplace, home, and future of aerospace. Their legendary past attracts aerospace specialists, scientists, engineers, and trainers to keep 'em flying faster, higher, farther, and safer than man has ever flown before. That made this the logical and perfect place for a faux laboratory."

"A false laboratory?" Marc spun in a slow circle, taking in the surroundings. The place looked pretty authentic to him. He glanced at Lei and could see she was puzzled, too. "I don't understand."

"When we couldn't find you," Mallory explained, "we created a laboratory and leaked that the U.S. was working on the final elements to make the Meissner Device reality."

"Why?"

"The plan," Thomas put in, "was that as long as your captors felt there was value to your life, they would keep you alive. That bought us more time to track you down."

"Which you never did," Marc noted.

Mallory sighed. "No one knows that better than me."

"Yet," Thomas said, "every day, we discovered new information related to the mole, the information leak, and your whereabouts."

Deverell opened an inner door and ushered the group into a highly technical facility. A half dozen men and women populated the space. In the center, under construction, was a larger model of the machine Marc had assembled in China.

"You put all this together for me?"

Mallory nodded. "We wanted to keep you alive until you could be located and brought home."

"A facade laboratory for a fake device," Marc said.

"Will you close this now?" Lei wanted to know.

Thomas shook his head. "As long as the Chinese and other foreign governments think we are working on this, they will continue to try to beat us to the formula."

Marc nodded. "Those Chinese will spend billions getting this going."

Lei looked sidelong at Marc. "They probably already have."

"Which deflects their funds from other pursuits of military interest," Mallory said.

Deverell unwrapped and shoved a stick of gum into his mouth. "Truth is, our government saw real potential in your design. Enough to assemble this facility for the development of the Meissner Device." He led them to a side room. Above the doorway was a sign that read, "Designgineering."

Inside, a heavy metal band played from one CAD system, fighting with classical music coming from a vinyl record player hooked up to a set of woofers and tweeters. Toys from the 50s, 60s, and 70s overflowed a toolbox. A slinky and an etch-a-sketch rested with dry erase markers in the whiteboard tray. The whiteboard and adjacent chalkboard wall were filled with equations and theorems. The place smelled of homemade banana bread.

Looking up from a set of blueprints, a man in Birkenstocks came to meet them, his body led by his belly. "You just missed the daily brainstorm session, but there is banana bread left." He indicated a tray that held slices topped with cream cheese, and one remaining poppy seed muffin.

"This is Wilson," Mallory introduced. "He is the design engineer for the project."

"Designgineer," Wilson corrected, shaking Marc's hand. Marc's eyes widened at the small patch on Wilson's shirt. Sitting up on its back legs, it was a skunk.

Marc recognized the official trademark for the Lockheed Martin Advanced Development Programs. Skunk Works projects were developed by a small, loosely structured group of experts who researched and developed purely for the sake of prototype innovation. The name was influenced by Al Capp's comic strip, *Li'l Abner*. Men and women with patches depicting skunks and crows were legends. Skunks were the elite engineers. Ravens were the electronics specialists. Old Crows were retired but couldn't keep away. They often showed up for the occasional consult or project.

Mallory went to the CAD, and over the heavy metal screams, tapped the user on the shoulder. He turned, and when he saw her, quickly turned down the music. Looking past Mallory, he locked eyes with Marc.

"Hebron?" Marc tried to take in the shock of seeing the redhead college student in this setting.

Then the boy was shaking his hand, a grin displaying a blue-black poppy seed stuck between his front teeth. "Mr. Wayne, good to see you."

"You're looking good, Hebron. Stronger." Marc flexed his fingers after Hebron's enthusiastic handshake and took in the boy's frame. "And you've filled out. I don't see those ribs anymore. Got rid of that tapeworm?"

"Had to let out my belt. This place has unlimited food, and this guy," he jerked a thumb to Wilson, "knows how to fill in during the between-meal lows."

"Made it myself." Wilson passed the plate to Marc.

To avoid a poppy seed in his teeth, Marc took a piece of bread. "What are you doing here?"

"Thought you knew." Hebron shrugged, his long arms still needing some sun. "Bringing your patent to reality."

From his pocket, Marc retrieved Hebron's diamond device. The boy brightened, and Marc returned the object to its owner. "I found another use—"

"For clarity of fine wine." Hebron opened his invention and peered inside. "I got your message from my professor."

"Something else, Hebron." Marc shook his head. "Your device has another use."

Lei came to Marc's side, watching the exchange.

"I used your invention to duplicate a badge that gave Lei and I unlimited access to restricted areas."

"That badge allowed us to find a way out." Lei put out her hand and shook Hebron's. "Thank you."

Chapter 35

Following a tour of the rest of the facility, Deverell brought the group back to the Designgineering room. "Everything is state-of-the-art."

"Except for Wilson," Mallory observed.

Deverell folded another stick of gum into his mouth. "Would you consider working with them on the blueprints?"

Marc took Lei's hand. "Is that an invitation, or a demand?"

Deverell grinned. "Hey, this is the United States of America. Land of the free, remember?"

"You choose if, and when, you visit," Thomas assured. "Maybe more importantly, you choose when you leave. We'll take you anytime we can get you and strictly on your terms."

Marc shook his head. "I'm really not interested."

"Everything is supplied," Thomas pressed. "You can have anything you want for your work. The most exciting aspect is that your patent for the Meissner Device is nearly operational."

"Except that troublesome spot," Deverell said. "Sir Galahad is still searching for the Holy Grail."

Again, Marc shook his head.

Mallory put a hand on his arm. "Don't make a decision now," she said gently. "Take your time. Think about it."

"Look," Deverell checked his watch, "let's get you both a good meal." He led the way back through security and outside where the Humvee they had arrived in was gone. In its place stood a bus usually designated to shuttle tourists from the main Wright-Patterson Air Museum to exhibits in outlying hangars. Climbing aboard, Marc was surprised to see they were the only passengers.

Once the party was seated, the driver swung the bus onto a main road. Passing the entrance, Marc noted the visitors making their way from the patriotic museum to the parking lot. Several stood for photos while others viewed the outdoor memorial. Open year round, the museum closed at 5:00 p.m. It was nearly 6:00 p.m.

Behind the Air Base's perimeter fence and west of the museum, the bus passed the Research and Development Flight Test Hangar and parked at the next structure.

Thomas led the group inside. Lined up in the quiet hangar like soldiers in formation, were row upon row of stately jets. Though they were different shapes and sizes, they all bore the Presidential Seal.

"Where are we?" Lei asked.

"I remember this. I came here for a school field trip." Marc inhaled deeply. "But it sure didn't smell this delicious."

The informal tour of the retired presidential aircraft gave Marc a renewed sense of security. He suspected that was the goal. Planes used by Franklin D. Roosevelt, Harry Truman, and Dwight D. Eisenhower sat like the sphinx, silent and proud. The centerpiece of the one-of-a-kind collection was a SAM 26000, used regularly by Presidents John F. Kennedy through Richard Nixon. During Nixon's shortened second term, the super plane served as the backup aircraft for the commander-in-chief. This was the plane that had winged President and Mrs. Kennedy to Dallas, Texas, on November 22, 1963, the day an assassin's bullet took the life of John F. Kennedy as Jacqueline Kennedy struggled in vain to hold her husband's brains inside his broken skull. Shortly after the assassination, Vice President Lyndon B. Johnson had been sworn in as president of the United States aboard the modified Boeing 707 that carried the body of the slain president back to Washington.

Deverell ran a hand over the sleek surface. "She's seen a lot, this plane."

"She wears it well," Thomas said. "A grand dame."

"The old girl got a face lift—a new paint job back in 2009," Deverell added. "Mirror, mirror on the wall, who's the fairest of them all?"

At the bottom of the jet's stairs stood a man smartly dressed in a black tux. He bowed slightly. "Welcome aboard. I will be your maître d'."

As they followed him up the gangplank, Mallory whispered to Marc. "He's one of ours."

The aroma of fine food grew stronger as the group entered the dining area of the luxurious plane that had once served the most influential men on the planet. An oval table was set for a seven-course meal. The smell of succulent roast prime rib came from the galley, and Marc's stomach growled.

"Lei, Marc, have a seat," Deverell said. "I've been looking forward to hearing your side of this crazy story."

While a waiter filled their goblets with ice water and topped each glass with a tangy slice of lemon, Mallory reached across and squeezed Marc's hand. "What a relief to have you here."

Marc lifted his glass in a toast. "To you heroes who got Lei and me back home."

"We're the same guys who got you into this adventure." Deverell flashed his "I-told-you-so" look to Thomas and Mallory.

Mallory looked sheepish. "I wasn't going to bring that up."

"It seemed like a good idea at the time." Thomas reached for a crab cake from the plate of appetizers.

Their server poured a red wine, and Marc thought of Hebron and his clarity invention.

"If the whole episode hadn't happened, I would still be back under Colonel Yao's control." Lei bowed slightly. "Being here is freedom for me."

"Here, here." Thomas lifted his wine glass in salute. "To freedom."

Tucking into a Caesar salad, topped with croutons and anchovies, Deverell turned the conversation. "Your Adi is an interesting character."

"His real life reads like something out of an action-adventure novel," Thomas put in.

"Maybe more like a contemporary suspense," Mallory said. "Or a mystery spy story."

Marc nodded. "He's something all right."

Lei leaned forward. "What do you know about him?"

Marc raised an eyebrow. "I thought you hated him."

Lei flushed. "I'm just curious."

Mallory put a gentle hand on Lei's arm. "Of course, you are curious after all you experienced in Israel." She gave Marc the look she used to tell him he was once again being a clueless male.

Deverell forked the last of his salad and pointed the dangling lettuce at Marc and Lei. "For Adi's safety, and the protection of his people and his work, everything about Adi must be kept secret."

"Just a kibbutz specializing in mud baths for tourists," Marc said.

"Adi works with illegal arms shipments."

"I figured that out myself," Marc told his sister.

Thomas dipped parmesan bread into herbed oil. "He receives and sends arms, then tips off the authorities about particular shipments that are dangerous to Israel or her allies. When the authorities bust those transports and pick up some nasty criminals, it looks like they were sloppy with their arrangements."

"Adi's work allows Israel to know who is doing what in their country," Deverell said. "And Israel is pivotal in world events."

"He was doing something to a shipment that went out just before he sent us," Marc recalled. "I'm no arms specialist, but from my shooting course in 4-H, it looked like he was tampering with the bullets."

The main course of honey-glazed carrots, steaming baked potatoes, and tender prime rib arrived. The last time Marc had tasted such a full-bodied American meal was his lunch with Mallory the day she flew into Indiana. A great deal had taken place since then.

"Brilliant actually," Thomas credited. "It's a trick borrowed from Vietnam. When American forces came across stockpiles of the enemy's weapons, they couldn't carry them out and certainly didn't want to leave the artillery for the enemy to use to kill Americans. They jimmied with some of the ammo, so the weapon would explode when fired by the first sorry fella who used it."

Deverell added horseradish sauce to his plate. "The same end will happen to the terrorists who fire the weapons in that shipment. The beauty of the plan is that the buyers assume the weapons are faulty from the factory."

Lei looked puzzled. "Adi is a good man pretending to be a bad man?"

"Exactly," Thomas said.

"It is dangerous work that Adi does," Lei murmured.

"Very treacherous." Mallory smiled her thanks to the waiter who cleared the table.

"Certain foreign governments are committed to obtaining the American trade secrets that can advance the development of their military capabilities," Deverell stated. "Foreign spying remains a serious threat."

"Adi is young to be making such world-impacting decisions," Marc mused.

"There's more." Mallory stirred cream into her coffee. "Adi is the son of the man who leaked your patent design for the Meissner Effect."

Chapter 36

"His father?" Lei deeply missed her father since his death. They shared a close and supportive father/daughter relationship, completely polarized from the absent father Adi had experienced. "Does Adi know?"

"Probably not," Deverell said. "They haven't been in contact for many years, not since Hatim Saad emigrated to the U.S. From the photo we found in Saad's apartment, he may have been blackmailed into trading information for his son's life."

The efficient waiter returned with a tray and placed before each diner a dish of chocolate mousse topped with blueberries and cream.

"Who are the blackmailers?" Marc stirred cream into his coffee. He leaned to Lei. "Look, beef and milk at the same meal."

Lei passed on the cream he offered. "Don't think I can mix the two anymore."

"We rounded up the woman inside the Secrecy Order office," Thomas said. "She was a sleeper."

"Sleeper?" Lei frowned at the unfamiliar phrase.

"Previously planted in a strategic position to be called into active service when needed at a later date." Thomas shook his head. "They can be a red herring for us. We've encountered spies who were adopted into the country and groomed for espionage. Others who are born here and cultivated for the same purpose."

"If the blackmailers used a photo of Adi to manipulate his father, that means someone knows about Adi," Marc said. "He's in danger."

Thomas spoke up. "Our conclusion is that whoever was blackmailing Saad knew Adi was involved in illegal arms shipments. Even the Israeli government knows that. The blackmail worked as long as the blackmailers believed Adi is who he pretends to be—a smuggler of illegal weapons."

"Or the blackmailers were threatening to share Adi's true motives with terrorists who buy from him," Marc said.

Their movements practiced and quiet, two waiters cleared the table. Balancing dishes stacked in precarious towers, one man carried out the plates while the second poured another round of coffee.

"We saw some of them. The terrorists." Lei shivered. "That may be worse for Adi and his people."

Marc recalled the brutes attempting to rape Lei. The man who swung his gun into Marc's head hadn't cared whether the blow knocked Marc unconscious or killed him. No wonder Reuven trusted no one. To protect their country, Adi and his team kept their enemies close.

"We've considered that option," Deverell assured. "Mossad has special agents keeping an eye on Adi for a while."

Bellies overfull, the guests lingered in the cushioned chairs around the presidential table. Long taut muscles in Marc's shoulders and neck were beginning to relax. Much remained to talk about and, he reminded himself, they were safe at home.

"Mossad?" Lei looked questioningly around the table.

"Those no-nonsense men that had helped the American spec ops team find us," Marc told her. "The American major had said they were Mossad."

"Israel's Institute for Intelligence and Special Operations." Thomas held up three fingers. "Mossad is one of three main entities in the Israeli Intelligence Community, along with Aman which is their military intelligence, and Shin Bet, which is internal security."

"And the kibbutz and his mother?" Lei asked.

"Yes. Keeping a protective shield on all of them." Deverell nodded. "Israel understands how important they are to freedom and security."

"His mother is quiet and stays in the background, but that merely keeps people from noticing her involvement," Marc said. "I can see where Adi gets his passion."

"Some things run in the family." Deverell stood and stretched. "Speaking of which, it's time to get you all back where you belong."

Mallory looped her arm around her brother's waist. "Now, you get home where you can rest over the weekend and return to work on Monday."

"Business as usual," Thomas added.

Marc rubbed his temples. "This adventure has all the elements of a classic espionage novel: the world teeters on the verge of World War Three, a foreign government focused on accessing our military secrets; foreign operatives who effectively use stealth and guile to gain that access; and an American government official who is willing to betray both her public office and the duty of loyalty expected from every American citizen."

"Don't forget the Holy Grail of weapons world powers are in a race to develop for military superiority." Thomas swept a hand toward Marc and Lei. "That's where you two come in."

"A contemporary suspense or mystery spy story," Mallory said.

"Or an action-adventure novel," Thomas put in.

"The hard part about saving my country and the American way of life as we know it, is not being able to brag about it to women," Marc moaned.

Thomas nodded knowingly. "Certainly, a drawback to the job."

“Hebron complained of the same ailment.” Mallory put her hands on her hips. “Would you really want a woman to love you for that?”

“Or for my superhero good looks.” Marc winked at his sister.

“And you.” Mallory looked to Lei.

Marc saw hope spring into the young scientist’s eyes. She had braved uncertainty, hardship, and danger to return to the United States.

“We’ve arranged for you to return to your college town.” Mallory smiled. “So, you can look up that young man of yours.”

Chapter 37

Marc Wayne pulled his gray streaked hair into a ponytail. He tugged on a jacket and threw a backpack over his shoulder. The ride downtown was short and chilly, past familiar brick houses in the neighborhood where he'd grown up. As he leisurely pedaled along the Midwest streets of Dixon, it felt good to be home. Secure. He didn't think he would ever again take that feeling for granted.

Wheeling around a corner, he biked down Main Street. Passing the bank, funeral home, and the Veterans of Foreign Wars Hall, he stopped in front of the former post office that now served as his patent office with his workshop in the back. Next to the door, the sign read *Marcus Wayne, Patent Attorney*. He unlocked the door and bent to pick up the newspaper.

"Mornin' Marc," came a gravelly voice.

Marc turned to see Dr. Thurmond standing outside his office door, peering at him over his glasses.

"Good morning yourself, Dr. Thurmond." He clasped the town vet's gnarled hand.

"It wasn't the same around here without you." Thurmond covered their handshake with his other hand. From the adjoining office space, Marc heard the insistent high-pitched yip of a small dog.

"How's business, Dr. Thurmond?"

"Barking along," the wizened old man quipped.

"Glad to hear that," Marc returned.

"I'm glad to hear anything at my age."

Marc smiled. "And just how old are you, Dr. Thurmond?"

"Old enough to remember when you used to come in here on your way home from school to play a game of checkers."

Marc tossed the newspaper inside where it landed on the receptionist's desk. "I was six."

Thurmond wagged a finger at him. "Looks like you're due for a round."

"An accurate diagnosis. What do you prescribe?"

The aged World War II veteran's eyes sparkled. "You be red. You always liked that color best. Best two outta three after work."

A raucous squawk emanated from the vet's open door. It sounded like the worn brakes on Adi's rusted four-wheel drive jeep. Marc recalled the day he and Lei met the Israeli man. After rescuing the illegal arms stowaways in the desert, Adi had driven them in his jeep. The jeep with the ear-splitting, semi-operational brakes.

"That doesn't sound good." Marc had commented diplomatically, clutching the grip above the passenger door handle while Adi careened along a steep hillside and dipped bumpily into a dry wadi bed.

"It's the desert." Adi shrugged. "No need for brakes."

The worn jeep had delivered Marc and Lei from the Negev to the safety of the kibbutz. The same vehicle that made its last, explosive desert trip as a decoy for the black-market weapons dealer intent on finding Marc and Lei and offering them back to Colonel Yao for a fine price.

Two more grinding squawks, more insistent than the first, brought Marc back to the present. "That doesn't sound like a dog."

"A parrot." Dr. Thurmond stuffed a fist into his pocket and retrieved something. Opening his palm, he exhibited a handful of sunflower seeds. In the shell. "Calling for more of these."

Marc stared at the small black and white striped shells lying loose in the old man's hand. "No kidding," his voice suddenly dry.

"What's that?" The man cupped a hand behind his ear.

Marc cleared his throat. "I said, no kidding."

Tossing a seed into the air, Dr. Thurmond caught it in his mouth. With his front teeth, he split the hull, and spit out the shell. "That bird has me trained. Got me eating these, too."

"It's easy to develop a taste for sunflower seeds." Marc was physically in Indiana, but his thoughts were back in Israel.

More insistent, the bird called again. "Stop in." The vet waved toward his clinic. "This bird is not something you see every day."

Marc fingered the package of sunflower seeds in his own pocket. "I'll do that."

He tossed a casual salute to his neighbor. In his office, he parked his bike next to the oversized pottery crock that served as an umbrella stand. That's when he remembered. The umbrella that Mr. Spencer would not leave behind. Marc had read of several designs that transformed umbrellas into shooting devices. From behind, Marc must have been injected with a sleeping agent. Colonel Yao seemed to be versed on such drugs.

Putting the past behind, Marc looked around his patent attorney office. As usual, he was there before Violet. Beside the receptionist's desk and under the window was a small table. The 8-track and stack of Elvis Presley, Lynn Anderson, and Charlie Daniel tapes had been replaced by a CD player. He thumbed through the short stack of CDs. Neil Diamond's classic album of *Hot August Night* Live at the Greek. Guns and Roses. Bon Jovi. The soundtrack from *Top Gun*. Violet had gotten adventuresome with her music.

He started down the hall, then turned back to pick up the newspaper from where he had thrown it on Violet's desk. Under the rubber band, the word China had caught his eye. He shook out the newsprint and scanned the front page.

"China Accused of Being Intellectual Vacuum Cleaner," the headline declared. "China deployed a diverse network of professional spies, students, and scientists according to a long-range plan to collect defense and industrial secrets," reported the head of counterintelligence for the Office of the Director of National Intelligence.

"This week, authorities arrested Kathleen Dongfan on espionage charges for passing classified information to China, and blackmailing employees in the United States Patent Office to supply designs classified Top Secret to foreign agents."

This must be the sleeper Deverell had mentioned over dinner. The USPTO employee that was involved in blackmailing Adi's father.

"A Cuban raised in Hawaii where her parents still live, Dongfan is one of a dozen investigations of Chinese espionage that have yielded guilty pleas in past months. The Immigration and Customs Enforcement officials have launched more than 540 investigations of aggressive illegal technology exports to China."

"Aggressive," Mark murmured. "That's an obscene understatement. Especially for those of us who wake up on the other side of the globe where someone sinister steals my ability to breathe."

"Recent prosecutions indicate that Chinese agents have infiltrated sensitive military programs pertaining to nuclear missiles, submarine propulsion technology, night-vision capabilities, and fighter pilot training. The critical information facilitates China's attempts to modernize its programs while developing countermeasures against advanced weapons systems used by the United States."

He thought of Mallory, Thomas, and the hyperactive, hyper thinking Deverell and realized that Mallory and her team had a big challenge ahead of them. "Today's prosecution demonstrates that foreign spy networks pose a grave danger to national security," said the assistant attorney general for national security. "We should all thank the investigators and prosecutors for effectively penetrating and dismantling this network before more sensitive information was compromised."

Rolling the paper, Marc snapped the rubber band back into place. Flipping the newspaper into the air, he caught it, flipped it higher and caught it again before leaving it on Violet's desk. Whistling, he started down the hall.

The phone rang, and Marc picked it up.

"Marc? This is Lei."

He dropped into his desk chair. "Lei. I didn't expect to hear from you so soon. How are you?"

"I wanted to thank you."

"Surely not for the splendid travel arrangements." Her light laughter delighted him, especially after previously seeing the stark fear in her eyes. "How's your Romeo?"

"Married and expecting his first baby."

The memory of her college co-ed had kept her hopeful for a new future while she was stuck in China. "I'm sorry, Lei. That must have been a shock." He thought back over her hardships to return to the man she loved. To the man she thought cared about her. The spunky electro-physics engineer conned powerful military minds on the opposite side of the globe, dodged despicable desert criminals, and concealed herself as contraband weapons to return to this man who once promised her his love. His loss. He obviously wasn't worthy of such a woman. "Are you all right?"

After a long silence, he asked again, "Lei, are you all right?"

He heard a tearful sigh. "That's a comparative question. All right compared to the conditions in the lab. Free from Colonel Yao's attentions. Liberated from the manipulations and stringencies of the People's Republic."

"How's your heart?"

She sniffed. "Broken."

"Understandable. I don't know any woman who worked as hard as you to reunite with the man she loved." He drew a question mark in the dust on his desk. "What will you do now?"

"I've come up with my own, as you say in America, half-baked idea."

"Go on." He heard her blow her nose.

"I'm thinking about applying to teach science at a university."

He grinned. "Of course. You'd be a natural. Will you invite me to guest lecture?"

"Is your passport valid?"

"Passport? Where are you applying?"

"Israel."

Marc rocked back in his chair. "Of all the places in the world, why Israel?"

There was a hint of excitement in her voice. "The more I think about our serendipitous encounter there… well… maybe I can—"

He stared out the window where his hometown was coming awake for the day. Two town councilmen followed an elderly farmer clad in overalls and a Carhart jacket inside the greasy spoon diner where the chef's special changed with the seasons and the coffee was all you could drink. "Sounds like you developed a taste for sunflower seeds."

She laughed. "In the shell."

"And hummus and eggplant."

"Dates and flat bread."

"Turkish coffee." Just the name made him check his teeth for coffee grounds.

"And an occasional mud bath."

"And maybe a zealous young man who introduced you to mud baths?"

Lei's reply was tenuous. And hopeful. "Maybe."

"Certainly intrigued. He is rather the superhero type of guy."

"Intrigued, yes."

"You'll be back in the center of action," he warned.

"If they can use me."

"I'd bet lunch that a university connection will come in handy. You'll let me know, won't you?"

"You'll be the first." He heard her take a breath and hesitantly begin. "Marc?"

"Mmmmm?"

"Would you think I'm crazy if I told you I had a dream that Adi kissed me?"

Marc gave a low whistle. "Funny you should ask."

"Why?"

"Because I had the same dream."

Chapter 38

In the middle of a clear night, Yao coasted off the unpaved road and parked. He sat for a moment and surveyed the surrounding area. Purposefully, he breathed slowly to steady the beat of his heart. Confident he was alone on the landscape, he picked up a package on the passenger seat and folded the thing into his pocket. He patted his other pocket, making certain the vital contents were neatly tucked inside.

On foot, Yao traced a direction that had been familiar to him as a boy. Both he and the countryside had altered since he had farmed alongside his father, and twice he lost his way and had to double back. In the cooling air, the warm earth smelled musky. A night creature skittered through the undergrowth, startling him.

At last, he spied the village. A collection of humble dwellings surrounded by fields that made up the whole of his childhood memories. Stealthily, he crept behind the finer home of the cadre, the man who enjoyed a higher level of living in exchange for being the probing, suspicious eyes of the People's Republic in this remote village.

Yao smirked his derision. The arrogant poppycock strutted his power among these placid families whose largest concern was to keep their bellies full. The cadre was no more than the fattest goldfish in a bowl, unaware of the great oceans teaming with rapacious predators.

Clouds blotted the starlight, and Yao stumbled. The noise disturbed the cadre's sleeping mutt who set up a racket and came running. From the package in his pocket, Yao pulled a dead rat and pitched the carcass to the mongrel. Like everyone else in the village except the fat cadre, even the dog never enjoyed the satisfaction of a satiated stomach. The unexpected meal consumed the animal's attention as the former resident knew would happen.

The dog whined warning growls to two other canines that approached, their noses high as they sniffed the air. Quickly, the three were circling each other, snapping, and tearing the rat between them.

With the noise of the dogs behind him, Yao stuck to the shadows but was less careful about being completely silent. Someone yelled at the dogs to be quiet, and Yao froze. The suspicious cadre peered into the darkness, looking for a long time in Yao's direction. Memories of the village cadre participating in the destruction of Yao's home and the beating of his parents flashed through his mind. Yao flexed into fists the fingers the cruel men had crushed under their boots.

Fighting erupted anew between the curs, and the cadre snapped at them to be quiet. Cursing, he kicked the mongrels apart and stomped back inside to his sleep.

A breeze blew, cooling the nervous sweat from Yao's face, and the clouds moved, unveiling a gibbous moon. He had chosen tonight gambling that the moon's glow had guided another to this same destination only hours earlier.

At the rear of his parents' modest home, he tapped lightly on the door. As he had hoped, someone had been waiting. Listening. The door opened immediately. His father's craggy face appeared, the old eyes squinted and blinked.

"Is it you?" The old man spoke like a dreamer not certain if he is awake or asleep.

Yao pressed the bottles into his father's hands. "I brought you medicine. For you and for mother's arthritis."

Then his father's bird-like arms were around Yao's neck, hugging him and drawing him inside. "My son. You are medicine enough."

Yao could smell the hearth fire and hope beat in his breast. He listened hard and thought he could hear murmuring. “Is he here?”

“Who, my son? Is who here?”

Yao took a deep breath. “The Brother.”

Chapter 39

In his workroom, Marc absorbed this new turn of events. He pictured Lei back in Israel, teaching students at the university. He knew she would be an unexpected and brave connection for Adi in his delicate balance between patriotism and the enemies of freedom.

Beside the stool at his workbench sat a sealed carton. He dropped his backpack and pulled out a pocketknife. With a slick slice from the blade, the carton sprung open to reveal a dozen bright red fire extinguishers. He grinned. The fire in his kitchen that had cost him the contents of a new fire extinguisher seemed years past. Definitely an event from a different, simpler life.

He recalled that early morning at home as he flushed ashes down the sink. His recipe for an adhesive had erupted into flames as the television on the kitchen counter reported the news that nations hostile to the United States had developed a superior jet engine. That same morning, Mallory had phoned, asking Marc to submit a patent application the FBI could trace through the United States Patent and Trademark Office. A simple request, so easy to do to help his sister with her important work guarding the security of the homeland. Such a simple decision when his world had been innocent.

In a short time, his life had been altered, and he knew there was no going back. The ill-fated experimental adhesive mixture had proved instrumental as an explosive, the first step in Marc's half-baked plan to get himself and Lei free from Colonel Yao's control. The first step in a series that landed them in a kibbutz deep in the Negev.

From his pocket, he retrieved the package of sunflower seeds and studied the rumpled cellophane label with Hebrew words.

"Well done, Adi," he said aloud. He propped the souvenir against his toolbox, snapped on the overhead light, and blew dust off the yellow legal pad that sat where he had left it on his workbench. Well, almost where he had left it. Someone had turned pages and written fresh calculations. Hebron, no doubt. Who else in Indiana, maybe in the world, had a grasp on the fantastic theory of the Meissner Effect.

That's where Violet found him later in the morning. Bent over new sketches and equations, reconciling Hebron's scribblings with what Marc had developed in China. Adding what he had picked up from Moshe in Israel. He knew she had arrived when the overhead lights came on. She must have seen his bicycle because he heard her surprised intake of breath all the way down the hall.

"Marc?"

He liked the way she said his name. He didn't bother to look up, suddenly aware of how hard his heart pounded at the sound of her voice. He tried to call back to her, but a surprising telltale catch lodged in his voice.

She hurried to the workroom. "Marc!" He heard warmth and joy in that single syllable. His name, spoken with deep affection.

He turned then and swallowed against the knot in his throat that told him volumes about his feelings for her. "How are you, Violet?"

Her smile was radiant. "I have a list of calls that came in while you were away. I've been docketing the due dates of communications from the Patent Office—"

He stood and put a finger to her lips. "But how are you?"

Her cheeks flushed. "Fine. Now that you're back." She smelled of spring lavender. "Where were you?"

He shook his head. "I would tell you if I could." She frowned, and he wanted to touch her forehead, to smooth away her confused emotions.

Violet stepped closer. "How are you?"

"Good to be back. Glad to see you."

She flushed deeper. "I'll make you some tea." She disappeared back down the hall.

He watched her go, heard her familiar movements in the small kitchen as she filled the pot with water and spooned fragrant leaves. In moments, the honest aroma of Earl Grey teased his nose. He thought of sweet Turkish coffee prepared by his Israeli hosts that he had quickly developed a taste for though he never felt comfortable with the thick layer of sediment lining the bottom of each tiny cup. Grounds that if he wasn't careful to leave in the cup, managed to stick in his teeth like Hebron's poppy seed and make him look like he needed dental work.

Steaming mug in hand, Violet returned and peered over his shoulder. "How's it coming?"

"In the tinkering stage but looking promising."

She set the cup near his elbow. "I see," she said, and they both knew she didn't. "All these inventions and you can't work the tea pot."

Marc smiled. "That's why I need you, Violet."

She regarded him. "Really?"

"Really." He looked intently at her.

"For making tea?"

He stood and cupped her chin in his hand. "For a lot more than making the perfect cup of tea."

Suddenly self-conscious, Marc dropped his hand. "Tell me how things are going with—what's his name—Hoss? Tom? Bob?"

"Rob."

"Right. I remembered that."

She tipped her head. "I can tell."

"What does he do, anyway? Lawyer? Doctor? Politician?"

"He builds and drives monster trucks."

Marc blinked. "Monster trucks."

"You know, the ones that show at ticketed events at the Fort Wayne Coliseum." She gestured to indicate something large. "The ones that make a lot of noise driving over the top of a long line of cars."

Unable to suppress the wonder, Marc laughed heartily.

"Are you making fun of him?" Violet's hand was on her hip like Mallory often did.

He put up his palms in surrender. "Quite the opposite. Living in a country where a guy can make a living driving monster trucks in public demolition demonstrations makes me want to salute something." He gently placed his hands on her shoulders. "I'm happy for you. Really."

Violet shook her head. "Rob and I are friends, Marc."

"Friends?"

She nodded. "That's all."

"That's wonderful." Butterflies suddenly danced in his belly.

"Wonderful?" The question reflected in her eyes.

"Really wonderful. Really, really wonderful." He bent and tenderly kissed her. He was lost in the surprising softness of her mouth when the phone rang. He felt her stiffen, knowing she thought she should answer the untimely interruption.

"That's the phone," she murmured, her breath sweet against his mouth. "I better get that."

"Who cares." He pulled her against him for a deeper kiss. As his arm moved to wrap around her waist, they both heard him bump the full mug. Instinctively, she pulled back to see the damage. Then she turned wide eyes back to his smiling ones.

Rather than crashing to the floor, the cup remained suspended in the air, as Marc would explain in due course, floating in a sea of magnetic fields.

Here's a sneak peek at the first chapters of the next Marc Wayne adventures, *Unnatural Cause.*

Unnatural Cause

Chapter 1

I should be a patient here somewhere, not standing upright, let alone walking.

Jeff Griggs gritted his teeth against the pain as he followed an older couple inside, close enough on their heels that he appeared to be with them, but enough behind that the twosome wouldn't react. The Oncology clinic waiting room in the Chicago Hospital should provide the anonymity needed for a few hours. The fact that they walked slowly proved an added bonus.

That bullet probably broke the ninth rib on my left side.

Broken ribs were a familiar companion in Griggs' life. The dang things were perpetually a magnet in his line of work. Good design, come to think of it. Or that bullet would have hit something important like a lung. Or his heart.

Last night may have been the first time a Tom Clancy novel saved a life. Luckily, the shot encountered the novel in an outer pocket of the vest. If not for the additional padding provided by the thick paperback, the Kevlar probably wouldn't have been enough. Plunking down cash for Clancy's highly detailed writing, plus the four-page foreword by the former United States joint chief-of-staff, had been the best $12.95 investment Griggs had made.

Down the hall and in sight of the rectangular waiting room with a bored looking receptionist, Griggs peeled off from his pseudo relatives in favor of the men's room. He glanced both ways as he entered. Good. Alone. But the door swung open a few seconds later, and he turned to the mirrors to catch a glimpse of the newcomer. In the mirror, he watched a hospital security officer follow him into the public restroom. He squelched a curse and bypassed the urinal. He must not make eye contact or do anything that would cause him to be memorable to the guard.

In a stall, he sat heavily. His ribs weren't the only area screaming agony. As best as he could tell, a ballistic tipped 44 magnum had caused this pain in his side. The guy knew how to place a shot but hadn't anticipated the quasi-fictional tome.

Nor the response from my SP-101 to his forehead.

The assailant's look of disbelief etched in Griggs' mind when the .357-jacketed hollow point did business with his gray matter. Griggs knew that would be the stuff of nightmares in the future. No matter how justified the shot, the image always came back to haunt him. Did anyone kill without being affected by the deed? Though he hated the spectral night replays that disturbed his sleep when he most needed rest, Griggs didn't ever want to be so calloused that the taking of a human life no longer stirred his conscience.

Even when the kill became necessary.

A troubling tidbit of that sub-portion of this nightmare remained in the fact that the dead man had been prepared—determined—to put a shot through Griggs' body armor. He nearly succeeded. It would have been game over.

Static, followed by a voice all business came from the security officer's radio. Griggs listened hard. Had an alert been received about him? Had the authorities initiated a search for anyone connected to the shooting? Once word got out, he wouldn't be able to take a dump without black-suited feds looking up his skirt.

"Be on the lookout for …"

Leaning so he could peer without being obvious, Griggs studied the security officer through the narrow opening between the locked door and the stall wall. Suddenly looking important, Barney Fife turned down the volume.

Griggs cocked his head, straining to hear the nasally dispatcher. "Searching … if you locate …"

Paul Blart keyed an acknowledgement, surveyed the room, and left.

So, the security officer would either search for Griggs—and anyone else connected with last night's escapade—or he had been sent on a hospital Easter egg hunt for a geriatric's lost teeth.

Chapter 2

"This guy is the poster child for eccentric inventors." Violet, smelling of lavender, stood in the doorway of his workroom.

Marc glanced up from the components he was fitting together. "Who?"

"Your first appointment of the day." She hooked a thumb over her shoulder. "He's either a clown in training with Barnum or his mother dresses him funny."

Marc's lab occupied the rear of the narrow building. Shelves along the walls held inventions in various stages of development. Perched on a stool at a cluttered table in the center of the room, Marc put a finger to his lips and nodded toward the opposite end of the table where the TV was tuned to the news.

Images of riots suddenly flashed across the screen of the television. "France continues to reel from the sudden death of French President Charpentier," the news anchor reported. "As conservative royalists claim the top leadership position, liberals chant reminders of the revolutionary guillotine."

When the newscast shifted to sports, Violet cleared her throat. "Marc. Your client."

"Yes." Marc used the remote to lower the volume. "You know, one of Einstein's housekeeper's jobs was to make certain the genius dressed before he left the house."

"Apparently, today is the housekeeper's day off." Violet inclined her head. "I put him and his nutty costume in your office."

Following Violet down the hallway, Marc wiped his hands on his pant legs. She glanced back pointedly at the smudges he deposited on his clothing.

"What?"

She rolled her eyes and stepped aside for him to precede her into the room where Marc wrote patent applications and met with clients. The appearance of his smallish personal office had certainly changed since Violet Seiwert had become his assistant. Though it had never been discussed during the interview nor outlined in the job description, she had easily infused style into the functional space. Antique fishing nets looped over a golf club hung horizontally framed the previously bare window. Opposite his broad wood desk, she had positioned a leather captain's chair to accommodate clients. But the chair sat empty.

Wearing a ridiculous wig, oversized clothing that sagged on one side, and dollar-store glasses, a man stood reading the plaques on the wall.

Congress shall have power
to promote the progress of science and the useful arts
by securing for limited times for their authors and
inventors the exclusive right to their respective writings
and discoveries.
— *Article 1, Section 8*
United States Constitution

Next to it, hung a second framed quote.

The question whether there is a patentable invention
is as fugitive, impalpable, wayward,
and as vague a phantoms as exists
in the whole paraphernalia of legal concepts.
If there be an issue more troublesome,
or more apt for litigation than this,
we are not aware of it.
— *U.S. Judge Learned Hand*
Supreme Court

Distracted by the amateurish disguise, Marc tried not to stare. Instead, he put out his hand to the visitor. "I am patent attorney Marc Wayne."

"Mr. David Jones." The man refused the handshake.

"Well then." Marc gestured for his guest to take a seat. "What can I do for you, Mr. Jones?" Long ago, the Midwest attorney learned not to dwell on the personalities of the clients he met in his professional life. In his experience, the more idiosyncratic the inventor, the grander the invention. However, today marked the first time someone had arrived with such outlandish theatrical preparation.

"May I have a glass of water, Mr. Wayne?" Not an unusual request, but the speaker delivered the words with a poor attempt at an accent.

"Certainly." Marc nodded to Violet, and she disappeared in the direction of the small kitchen.

Marc took his seat on the opposite side of the eighty-year-old walnut desk built by his grandfather from timber harvested at the family farm in Northern Indiana.

His guest leaned forward. "Is this session being recorded?"

Violet returned and took her time removing Marc's nearly empty mug of cold Earl Gray tea and arranging a water glass on a ceramic coaster near Mr. Jones.

Marc shook his head. "Should I record it?"

"Absolutely not!" Mr. Jones looked pointedly at Violet.

Marc cleared his throat. "Violet, please close the door on your way out."

The door closed with a click just enough louder than usual for Marc to know Violet would rather have stayed. She shared his curiosity about the mysterious visitor.

Marc began with his usual introduction. "I have a few initial questions to assure your subject matter will not be in conflict with my other clients—"

"We will begin with a $50,000 retainer." From an oversized overcoat pocket, the strange man produced two large stacks of bills held together with rubber bands. He thumped the cash on the desk.

Begin? Patent attorney and inventor, Marc Wayne, estimated the stack of one-hundred-dollar bills on his desk measured two inches thick. *$50,000 initiated a beginning? Beginning of what?*

From another pocket, the client produced a paper he set before Marc. “It is essential that we establish attorney-client relationship. Now, once you sign this contract, we will proceed.”

A glance through the wording confirmed a lawyer had prepared the document. Signing an agreement for a privilege already granted to the client felt strange to Marc. “Confidential communications between attorney and client is one of the oldest legal concepts. Unless the information you disclose is used to further a crime, tort, or fraud, I can assure you—”

“I insist.” The man tapped the signature line with an index finger missing a fingernail. “Before I make full and frank disclosures, you must sign this contract binding you to confidentiality.”

Marc had seen this scenario countless times; rabid to protect their baby, inventors sought stringent, even waterproof legal protection. Though he had yet to encounter a hypothesis that warranted such precautions. Besides, an attorney that violated this concept could be disbarred. Regardless, inventors seeking their first patent were frequently paranoid about someone stealing their brilliant idea.

Studying his visitor, Marc could see that no amount of reassurance would convince the man that such an agreement wasn’t necessary. This stranger had taken paranoia to a new level.

“Your signature, please.”

Rereading the words, Marc ascertained that the agreement reiterated the common understanding of attorney-client privilege. He picked up a pen and wrote his name.

“Very well.” The paper disappeared back into Mr. Jones’ oversized overcoat pocket. “Let me be upfront. I understand that your sister, Mallory Wayne, is a Special Agent of the FBI, and the two of you were in the news not long ago regarding an invention of yours stolen by the Chinese. Because of your connection with her, and your familiarity with technology, I have come."

His rehearsed speech delivered, Mr. Jones pulled what appeared to be a handheld GPS from his shirt pocket.

Why did Mallory interest this peculiar person? Before Marc could ask, Mr. Jones set the device near the glass and pressed a button.

Did the mechanism determine some attribute about the glass of water? Marc quickly theorized a number of possibilities. But the water and the gadget remained motionless.

Mr. Jones broke the momentary silence. "You see, I need to bring current events to the table so that I might explain what is needed." He held up a finger for each of the three names. "The death of a supreme court justice two months ago, the death of a senator two days later, and the death four days ago of the French President."

The curious visitor slipped the device back into his pocket. "None of these men were old, in fact, they ranged from forty to fifty-five years of age. Do you know the causes of death?"

Marc recalled the television news report about the French President he had viewed just before meeting Mr. Jones. As Marc mentally tried to connect Mallory, the glass of water, and these governmental leaders, Mr. Jones answered his own question. "Embolism. A small, but deadly air bubble in the bloodstream."

Glancing back to the glass, Marc noted the water remained quiet.

"What is not widely known," Mr. Jones lowered his voice conspiratorially, "is they were all gas embolisms, and the deaths each occurred in public places the men were known to frequent. These precise facts have raised suspicions."

Marc tossed out a cliché. "I have heard it said that deaths seem to occur in threes."

Mr. Jones gave a wry smirk. "Scientific minds don't entertain superstitions." Like a magician about to execute some sleight of hand, the man retrieved the GPS-like unit from his pocket. "I will provide a few details to end your puzzlement but now a demonstration. Observe the glass of water. I have neither touched nor put anything into it."

He pressed buttons. A bubble appeared in the water and rose to the surface. Reaching the air, the bubble popped, spewing water droplets onto the antique desk's polished surface.

A simple bubble in a glass of water suddenly turned profound. The possibilities for evil flooded Marc's mind. Here was the ability to end a life in a fashion that appeared natural. The ramifications were staggering. World power could be manipulated by the simple pressing of buttons.

The holder of this device had the ability to kill without suspicion. To commit murder without consequence.

Chapter 3

With the security guard gone, Griggs emerged from the bathroom stall. At the sink he lathered up and gave his face, hands, and arms a scrubbing. Washing off as much of the night as he could, he observed his surroundings.

The fourth stall had been occupied when Griggs entered. He could hear heavy breathing behind the stall door, and then a netbook was placed on the floor. Silently, Griggs approached, and stealthy as a cat, reached under the door and pilfered the laptop from a guy who thought his porn viewing in a stall would go unnoticed.

Exiting the restroom and passing through the reception area, he picked up a bottle of water under a sign that read "Drink to your health." In a corner facing the entrance door, Griggs took a seat. The hallway to his immediate left might provide an escape route if needed. Was that his training kicking in? Or paranoia?

He downed the water. The fog in his mind started to clear. Someone who knew Griggs liked Barq's root beer had been added to his enemy list. Whatever substance he ingested had almost certainly been in the opened bottle of root beer that he had found in the fridge while babysitting the FMOTUS last evening. The phrase, First Man of the United States, never caught on when Flo Garcia became president, but the Secret Service used the acronym anyway.

Was Griggs still a member of the Secret Service? After last night's events, he couldn't be certain. The netbook flickered to life, and an image appeared on the screen. At least the handyman in the restroom favored adult women. The internet connection powered up, and Griggs went to gmail and established a new account. *Westminster2435@gmail.com*. Hopefully, Jeremy would recognize the street and house number of the foster home where he and

Griggs met. The email must catch his eye, or Jeremy would simply relegate the message to spam.

A television in the hospital reception area entertained patients and distracted them from the ridiculous disparity between the scheduled appointment and when the nurse eventually invited the patient back to see the doctor. They called the space a waiting room for a reason. Griggs listened to the flat screen, tuned to the news channel, for any report featuring last night's activities.

For an additional precaution, he opened a window on the netbook to Internet news and felt relieved that he was not the headliner. Periodically scanning the people coming and going through the oncology department, Griggs began his email.

Jeremy,

J.G. here. I need to make a record about what is happening and don't know who to trust but you. From what I see on the internet and television, none of what happened last night here in Chicago is in the news yet. I have to assume that someone or some group will spin this shortly, and I may be the fall guy.

First, I presuppose the news has not surfaced because the open ends are not yet dealt with. Meaning that I am still alive. The other possibility is that they have not ascertained the whereabouts of my charge. I can't use his name—the NSA guys may be scouring emails for his unique handle. Yesterday, I was assigned babysitter of Pres. G's husband.

I now have a broken or badly bruised ninth rib—you would say, a vertebrochondral rib—caused by either an assassination or kidnapping attempt. I don't even know if the guy is still alive.

Briefly, here is what happened. Orders yesterday morning were to drop everything to tend to my new assignment and make sure he got what he wanted, which I know is not exactly the same way you meet needs in your missionary hospital. Toward evening, I learned he is partial to pixie brunettes and a meeting had been arranged at a safe house in a quiet neighborhood. My super told me to make sure we were out of sight, and no one should know. Just Baby, me, and my partner, Joe.

Someone dropped off Pixie ahead of our scheduled arrival, in fact, I am not sure if she was in on this or not. Anyway, she doesn't have anything to say now, but I am skipping ahead.

I waited in the kitchen with Barq's while Pixie and Baby were getting cozy. Joe went outside for a smoke and to take a perimeter walk. Sometime after, I felt like a mule kicked me in the side, and I slammed into the fridge and hit the floor behind the island.

The guy that came in the door acted a bit overconfident. He didn't realize I was a Clancy fan, and the Barq's tasted flat, (I'll explain later). I gave him something to think about from my revolver.

My head spun, and I slowly got up from the kitchen floor. I am speculating what happened next. Pixie must have run to the front door, whether part of the plan or to escape, I don't know. At that moment, what sounded like two AK-47s put holes in the front of the house, and Pixie clearly had no body armor to slow the rounds that hit her.

Griggs shifted his weight to ease the pain in his side. *Had Pixie's death been part of the plan? Or a backup plan?* He checked the news again for updates, relieved that neither his face nor the face of the president's husband made the headlines. A quick scan of his surroundings assured that everything appeared to be usual business protocol. He turned his attention back to the message for Jeremy.

Possibly the guys in front were discerning enough to know that the last shot did not come from their friend's cannon. When I got to the living room, Baby had passed out on the couch, but otherwise seemed untouched. Squealing tires led me to believe that the fireworks were over.

The fog in my mind had slowed its spinning, and I thought of Joe. Outside, I found him on the ground next to the car. Blood poured from a knife wound precisely aimed beneath his vest to reach his lungs. I knelt close in time to hear his last gurgling breath.

Both my body and mind were dazed, and the only thing I could think was to get out of there. I got Baby into the car, but then thought what if the bad guys had another backup plan. I looked under the car and saw something flat and small attached to the gas tank. Whether a tracking device or an explosive with a trembler switch igniter, I didn't stick around to mess with it.

I half carried and half walked Baby down the street to a church where the side of the building shielded us from streetlights. The sirens grew louder, I got the old church van started, and Baby and I were going down the street. Apparently under the influence of a root beer borne drug, I wrecked the van. Remind me someday to donate money to the Second Street First Baptist Church. For a new van.

My only clear thought was that I was on my own. No one to trust. And leaving Baby in an open public place would get him attention that should serve to protect him. I was a solitary man with no one to love (sorry Neil).

We made it to the "L" and got settled. Most passengers must have thought we were a couple of late night/early morning drunks, so we were ignored. I put sunglasses and a hat on Baby and tucked a newspaper under his arm.

At the next stop, I got off the "L." I left Baby behind, still snoozing, to be discovered in his apparently inebriated state.

What tale is being woven now to put this on the outside guy (me), I have no clue. No doubt the story will include a silenced ending for me.

Gotta go, thanks for helping me think this through.

Later,

J. G.

Chapter 4

Marc stared across his desk at the costumed man.

Slipping the device—or was it a weapon—back into his pocket, Mr. Jones held Marc's gaze. "Now you are wondering if I am a murdering monster."

Marc had indeed been considering a laundry list of options. "It seems unlikely someone with such political intrigue as a murderer of three high profile men would come to me with a confessional."

"Let me tell you a bit about what you just saw and what we—er—I want." Mr. Jones faltered as if aware the slip of tongue was regretted but could not be undone. "A weakness with most weapons directed against a person or target is they typically have a predictable path. Countermeasure weapons track the source. A muzzle flash or the route of the weapon betrays the location of the starting place, so the potential danger may be eliminated."

Marc rotated the glass and studied the water. "Are you saying what I saw does not have a source?"

"It does." With dramatic flair, Jones reached for the glass and drank half the contents. "And it doesn't. As you know, electromagnetic energy permeates us and the space around all the time, whether it be the radio signals from the inane country music station on the edge of a small town or the superior energy from the sun." He paused. "Are you following me so far?"

Marc nodded. "Go on."

Mr. Jones picked up the water glass again. "All that has been done to the water is that a small amount of energy was focused within the glass." He met Marc's eyes. "The path from which the energy came was quite literally from a multitude of directions."

"How much energy?"

"Not a lot." He handed the glass to Marc. "You see, the water has not even noticeably warmed. But this focused energy was so concentrated at one point that it turned a small amount of the water—"

"Into a steam bubble." Marc tipped the glass first one way and then the other. There was no discernable temperature difference to the glass, nor did the water appear or smell altered. "Your handheld mechanism is not the source. It merely relayed coordinate information to the tool that caused this?"

Jones gave a slight smile of acknowledgement. "My device profiled the energy response of this space close to the water with an extremely low-level spread spectrum electromagnetic signal and calculated reflected signals in terms of angles and time delays." The wig dipped low over his left eye, and he pushed the hairpiece back into place. "Energy could then be subsequently focused in many directions to cause a diminutive—"

"But lethal—"

He tipped his head. "But lethal—bubble in your glass. Unlike this simple glass of water, complex humans are actually quite fragile creatures, and if a small bubble suddenly occurs in the blood…"

"The result is death."

"Let me sum this up, as I have already misspoken and revealed that I am not alone in this endeavor." Jones shifted uneasily in his chair. "We discovered this phenomenon and perfected this device several years ago with the hope of using it in a field to benefit mankind."

Marc's eyebrows shot up. "How could creating embolisms benefit anyone?"

"And," Jones tapped the device in his pocket, "we quickly realized this could be used as a nearly undetectable instrument of death. So, rather than making it public or applying for a patent or showing the military, we simply locked away the device."

"But the three deaths?"

"Precisely." The visitor nodded. "The circumstances of these deaths raised suspicions that someone has developed similar technology or worse, discovered our work. Particularly, since each deadly embolism occurred in a place easily profiled, just as I profiled the location of the glass of water."

Marc replaced the glass on the coaster. "And you want my sister, Mallory, to inquire into these events?"

"To stop whoever employs this dangerous power for political gain."

"But you, of course, have no political interests."

Mr. Jones slammed his fist on the desk. "We are not political." He took a calming breath and lowered his voice. "We want the authorities to be aware of what can be happening and do their job. But we insist on complete anonymity."

Marc gestured at the lopsided wig. "Hence your disguise?"

Jones adjusted his glasses. "I apologize for the quality, but we do not want to be forced to be involved."

Leaning forward, Marc rested his arms on his desk. "What exactly do you want me to do? Tell my sister what you showed me? Do you believe this fantastic story will send the FBI hounds out sniffing for the perpetrators?"

"A concern we anticipated." Mr. Jones looked at his watch. "It is presently 9:47 a.m. Exactly one week from now, a similar energy pulse will be directed four inches above that coaster on your desk. You may put any target there that you choose and have whomever you want here to observe. You can use a glass of water." He rose to leave. "Or to convince your guests, use a biological target."

As Marc stood, Mr. Jones ordered, "Sit!"

Marc slowly sank back to his chair.

“Recall that I insist on complete anonymity.” He patted his overcoat pocket that held the legal contract. “Do not go to the windows or allow your secretary to leave the building.”

“Mr. Jones, this is hardly necessary—”

The man in the disguise waved him away. “I am serious about not being discovered and have profiled several coordinates in your office.”

Marc remembered taking several minutes to listen to the news report in his lab while Mr. Jones was alone in his office. “I understand.”

At the door, Jones turned back. “I also profiled coordinates in your secretary’s office.”

Marc must have shown concern because Jones smiled slyly. “I see I have your attention. If I detect that I am being watched or followed, I will activate the device to randomly pulse those coordinates.”

The thought of Violet dropping dead made Marc want to slam a fist in his visitor’s face. Despite his claim to have locked away the technology, Mr. Jones appeared more than willing to push lethal buttons today.

Jones opened the door. “Now, call your attractive secretary and keep her here in your office for fifteen minutes.”

To read more, get a copy of *Unnatural Cause* by P.S. Wells and Max Garwood.

Thank you for reading *Secrecy Order,* Book 2 in the Marc Wayne Adventure series by P.S. Wells and Max Garwood.

If you have a moment, please leave a review on your favorite bookseller website. Reviews are the best gift you give an author.

If you haven't read the first book in the Marc Wayne Adventure series, *The Patent,* pick it up at your favorite bookseller.

The prequel to *The Patent* is *Chasing Sunrise* by P.S. Wells, available in e-book, paperback, and in a dramatic audio version.

Other titles you may like by P.S. Wells include:

- *Chasing Sunrise*
- *Homeless for the Holidays*
- Check out the audio version of *Homeless for the Holidays* read by voice actress Katie Leigh
- *The Ten Best Decisions A Single Mom Can Make*
- *Slavery in the Land of the Free*
- *The Girl Who Wore Freedom*

Connect with P.S. Wells at www.PeggySueWells.com. I'd like to hear from you.

Made in United States
North Haven, CT
20 October 2026

20740132R00095